III-Nitride Materials: Properties, Growth, and Applications

III-Nitride Materials: Properties, Growth, and Applications

Editors

Yangfeng Li
Zeyu Liu
Mingzeng Peng
Yang Wang
Yang Jiang
Yuanpeng Wu

Basel • Beijing • Wuhan • Barcelona • Belgrade • Novi Sad • Cluj • Manchester

Editors

Yangfeng Li
College of Semiconductors
(College of Integrated
Circuits)
Hunan University
Changsha
China

Zeyu Liu
School of Physics and
Electronics
Hunan University
Changsha
China

Mingzeng Peng
School of Mathematics and
Physics
University of Science and
Technology Beijing
Beijing
China

Yang Wang
Songshan Lake Materials
Laboratory
Dongguan
China

Yang Jiang
Institute of Physics
Chinese Academy of Sciences
Beijing
China

Yuanpeng Wu
Department of Electrical
Engineering and Computer
Science
University of Michigan
Ann Arbor
USA

Editorial Office
MDPI AG
Grosspeteranlage 5
4052 Basel, Switzerland

This is a reprint of articles from the Special Issue published online in the open access journal *Crystals* (ISSN 2073-4352) (available at: https://www.mdpi.com/journal/crystals/special_issues/MN5BE57M2J).

For citation purposes, cite each article independently as indicated on the article page online and as indicated below:

Lastname, A.A.; Lastname, B.B. Article Title. *Journal Name* **Year**, *Volume Number*, Page Range.

ISBN 978-3-7258-2107-5 (Hbk)
ISBN 978-3-7258-2108-2 (PDF)
doi.org/10.3390/books978-3-7258-2108-2

© 2024 by the authors. Articles in this book are Open Access and distributed under the Creative Commons Attribution (CC BY) license. The book as a whole is distributed by MDPI under the terms and conditions of the Creative Commons Attribution-NonCommercial-NoDerivs (CC BY-NC-ND) license.

Contents

Preface vii

Yangfeng Li
III-Nitride Materials: Properties, Growth, and Applications
Reprinted from: *Crystals* 2024, 14, 390, doi:10.3390/cryst14050390 1

Zizheng Li, Huimin Lu, Jianping Wang, Yifan Zhu, Tongjun Yu and Yucheng Tian
Maximizing the Light Extraction Efficiency for AlGaN-Based DUV-LEDs with Two Optimally Designed Surface Structures under the Guidance of PSO
Reprinted from: *Crystals* 2022, 12, 1700, doi:10.3390/cryst12121700 7

Zhenyu Chen, Feng Liang, Degang Zhao, Jing Yang, Ping Chen and Desheng Jiang
Investigation into the MOCVD Growth and Optical Properties of InGaN/GaN Quantum Wells by Modulating NH_3 Flux
Reprinted from: *Crystals* 2023, 13, 127, doi:10.3390/ cryst13010127 16

Feng Tian, Delin Kong, Peng Qiu, Heng Liu, Xiaoli Zhu, Huiyun Wei, et al.
Polarization Modulation on Charge Transfer and Band Structures of GaN/MoS_2 Polar Heterojunctions
Reprinted from: *Crystals* 2023, 13, 563, doi:10.3390/cryst13040563 27

Atse Julien Eric N'Dohi, Camille Sonneville, Soufiane Saidi, Thi Huong Ngo, Philippe De Mierry, Eric Frayssinet, et al.
Micro-Raman Spectroscopy Study of Vertical GaN Schottky Diode
Reprinted from: *Crystals* 2023, 13, 713, doi:10.3390/cryst13050713 39

Yujian Zhang, Guojian Ding, Fangzhou Wang, Ping Yu, Qi Feng, Cheng Yu, et al.
Normally-Off p-GaN Gate High-Electron-Mobility Transistors with the Air-Bridge Source-Connection Fabricated Using the Direct Laser Writing Grayscale Photolithography Technology
Reprinted from: *Crystals* 2023, 13, 815, doi:10.3390/cryst13050815 51

Jiaping Guo, Weiye Liu, Ding Ding, Xinhui Tan, Wei Zhang, Lili Han, et al.
Analysis of Photo-Generated Carrier Escape in Multiple Quantum Wells
Reprinted from: *Crystals* 2023, 13, 834, doi:10.3390/cryst13050834 63

Sanjie Liu, Yangfeng Li, Jiayou Tao, Ruifan Tang and Xinhe Zheng
Structural, Surface, and Optical Properties of AlN Thin Films Grown on Different Substrates by PEALD
Reprinted from: *Crystals* 2023, 13, 910, doi:10.3390/cryst13060910 71

Swarnav Mukhopadhyay, Cheng Liu, Jiahao Chen, Md Tahmidul Alam, Surjava Sanyal, Ruixin Bai, et al.
Crack-Free High-Composition (>35%) Thick-Barrier (>30 nm) AlGaN/AlN/GaN High-Electron-Mobility Transistor on Sapphire with Low Sheet Resistance (<250 $\Omega/\square$)
Reprinted from: *Crystals* 2023, 13, 1456, doi:10.3390/cryst13101456. 82

Swarnav Mukhopadhyay, Surjava Sanyal, Guangying Wang, Chirag Gupta and Shubhra S. Pasayat
First Demonstration of Extrinsic C-Doped Semi-Insulating N-Polar GaN Using Propane Precursor Grown on Miscut Sapphire Substrate by MOCVD
Reprinted from: *Crystals* 2023, 13, 1457, doi:10.3390/cryst13101457 95

Preface

III-nitrides have been widely developed and researched over the past 30 years. Gallium nitride (GaN)-based light-emitting diodes (LEDs) play an important role in lighting, display, light communication, sterilization, etc. Compared to blue and green LEDs, GaN-based yellow or red LEDs and ultraviolet LEDs are deficient in high external quantum efficiency; however, striking progress has been made in recent years. GaN-based high electron mobility transistors (HEMTs) have already shown tremendous potential for high-frequency communications and power conversion. In comparison, the relatively high defects in the epilayers and interfacial traps hinder the practical performance of GaN HEMTs (e.g., breakdown voltage, Vth hysteresis, high dynamic Ron, etc.). The p-channel field effect transistors (FETs) of GaN are also under investigation. Compared to its incumbent Ga-polar counterpart, N-polar GaN demonstrates some intrinsic merits for both LEDs and HEMTs. Therefore, studies on N-polar GaN are also flourishing. Regarding GaN growth, metal–organic chemical vapor deposition (MOCVD), molecular beam epitaxy (MBE), and hydride vapor phase epitaxy (HVPE) constitute the major methods employed; however, other methods such as plasma-enhanced atomic layer deposition have also demonstrated exciting achievements. The growth kinetics of GaN still remain to be clarified, especially on foreign substrates (e.g., sapphire, silicon, silicon carbide, etc.). Many open questions regarding III-nitrides await the consensus of researchers.

The following Special Issue (SI) includes 12 excellent works, including 10 research articles and 2 reviews. The SI covers a plethora of topics on III-nitride materials and related devices in frontier fields including structural design, growth methods, device fabrication technology, investigation of physical properties and mechanisms, etc.

We hope that this Special Issue will contribute to the promotion of the exploration of III-nitride materials and applications.

We highly appreciate the transcendental research work of all authors. We are also indebted to Mr. Mars Tan from the Editorial Office for his continuous assistance and full support.

Yangfeng Li, Zeyu Liu, Mingzeng Peng, Yang Wang, Yang Jiang, and Yuanpeng Wu
Editors

Editorial

III-Nitride Materials: Properties, Growth, and Applications

Yangfeng Li

Changsha Semiconductor Technology and Application Innovation Research Institute, College of Semiconductors (College of Integrated Circuits), Hunan University, Changsha 410082, China; liyangfeng12@mails.ucas.ac.cn

1. Introduction

Since the activation of magnesium (Mg) in p-type gallium nitride (GaN) [1,2], striking progress has been made in III-nitride materials in terms of properties, growth, and applications [3]. Nowadays, aluminum nitride (AlN) epitaxially grown on nano-patterned AlN/sapphire template has a lower dislocation density as low as 3.3×10^4 cm^{-2} [4–6]. The behavior of impurities such as carbon in GaN has also been thoroughly investigated [7]. Recently, methods for characterizing the edge dislocation density of a thin film [8] or the interface roughness of multiple quantum wells (MQWs) or superlattice structures have also been further developed [9]. In addition, the light emission mechanisms of InGaN have also been investigated through the clarification of localized states [10] and the direct observation of carrier transportation between different localized states [11,12]. Some mechanisms still need to be unveiled, such as the abnormal enhancement of photoluminescence (PL) intensity in the mid-temperature range of InGaN materials during the temperature-dependent photoluminescence (TDPL) measurement [13–18].

Due to their high luminous efficiency and widely tunable bandgap, InGaN light-emitting diodes (LEDs) have permeated our daily lives. The properties of InGaN LEDs are still being improved, such as the light-output power, the lower leakage current, the efficiency of green, yellow, orange, red, and ultraviolet LEDs, etc. [19–27]. The external quantum efficiency (EQE) of InGaN blue LEDs is over 80% [28], while that of green ones surpasses 50% [29,30]. Great breakthroughs are also made in the "green gap" range [31]. Novel treatments of the quantum barrier have also rendered thrilling properties [32–35]. Single-chip white light has also been investigated [36]. In recent years, with their merits of high brightness, fast response, and high resolution, microLEDs have played an important role in next-generation displays such as augmented reality, full-color matrix automotive lamps, pico-projectors, etc. [30,37–42]. The EQE of InGaN red microLEDs has already reached 7.4%, demonstrating promise for application soon [43,44]. Blue and green InGaN lasers have also achieved inspiring performances [45,46]. Photonic integrated circuits with lower power consumption may alleviate the heat dissipation problem in computers. Microdisk lasers are promising light sources in photonic integrated circuits [47]. The threshold, quality, and other properties are gradually being improved [47–52].

Owing to the polarization effect, two-dimensional electron gas (2DEG) can be generated adjacent to the AlGaN/GaN interface, thus facilitating the development of high-electron-mobility transistors (HEMTs) [53]. The enhancement-mode (i.e., normally off) GaN HEMT has been demonstrated by fluoride-based plasma treatment for the first time [54]. Nowadays, the p-GaN gate is more commonly used to obtain enhancement-mode GaN HEMTs [55]. At present, p-channel GaN field effect transistors (FETs) are attracting attention, with potential in fabricating GaN-based complementary integrated circuits [56–59].

Benefiting from the inversed polarization field, N-polar GaN may surmount the bottlenecks faced by their incumbent Ga-polar counterparts [60,61]. Achievements have been made in the metal–organic chemical vapor deposition (MOCVD) growth of N-polar GaN [62–66]. Additionally, HEMTs based on N-polar GaN have exhibited transcendental performance in some aspects [67–69].

Citation: Li, Y. III-Nitride Materials: Properties, Growth, and Applications. *Crystals* **2024**, *14*, 390. https://doi.org/10.3390/cryst14050390

Received: 10 April 2024
Accepted: 22 April 2024
Published: 23 April 2024

Copyright: © 2024 by the author. Licensee MDPI, Basel, Switzerland. This article is an open access article distributed under the terms and conditions of the Creative Commons Attribution (CC BY) license (https://creativecommons.org/licenses/by/4.0/).

This Special Issue has collected recent research focused on the properties, growth, and applications of III-nitride materials. It contains ten articles and two reviews, which will be briefly described in the following paragraphs.

2. An Overview of Published Articles

Li et al.'s article (contribution 1) used an intelligent algorithm to investigate the light extraction surface structure for deep-ultraviolet LEDs (DUV-LEDs). As a result, compared to conventional structures, the optimized truncated pyramid array (TPA) and truncated cone array (TCA) structures enhanced the light extraction efficiency (LEE) of the DUV LED with an emission wavelength of 280 nm by 221% and 257%, respectively.

Chen et al. (contribution 2) investigated the surface evolution and emission properties of InGaN/GaN MQWs with different ammonia flow rates by MOCVD. Different ammonia flow rates led to different temperature-dependent photoluminescence (TDPL) behaviors. Combined with atomic force microscopy (AFM), the underlying mechanisms were investigated in detail.

The third article (contribution 3) was composed by Tian et al., who investigated GaN/MoS$_2$ heterostructures by first-principles calculations. The heterojunctions of N-polarity GaN/MoS$_2$ and Ga-polarity GaN/MoS$_2$ are compared from the binding energy. A type-II energy band alignment occurs between both the Ga-polarity and N-polarity GaN/MoS$_2$ polar heterojunctions, although the directions of the built-in electric field are opposite. Moreover, the energy band alignment could change from type II to type I by exerting in-plane biaxial strains on GaN/MoS$_2$ heterostructures.

The fourth article (contribution 4) by N'Dohi et al. investigated the physical and electrical properties of vertical GaN Schottky diodes. The correlation between the reverse leakage current and doping, as well as dislocations, was investigated.

Zhang et al. (contribution 5) fabricated a normally off p-GaN gate HEMT with an air-bridge source connection. The as-fabricated HEMT exhibited an on-resistance as low as 36 $\Omega \cdot$m, a threshold voltage of 1.8 V, a maximum drain current of 240 mA/mm, and a breakdown voltage of 715 V.

In the sixth article (contribution 6), Guo et al. quantitatively analyzed the relationship between carrier energy and the potential height to be surmounted in the GaAs/InGaAs MQW structure by considering the Heisenberg uncertainty principle. Pump–probe technology is adopted to determine the lifetime of the photo-generated carriers under short circuit (SC) and open circuit (OC) conditions.

Liu et al. (contribution 7) employed plasma-enhanced atomic layer deposition (PEALD) to grow aluminum nitride (AlN) thin films on Si (100), Si (111), and c-plane sapphire substrates at a low temperature of 250 °C. The as-grown polycrystalline AlN thin films had a hexagonal wurtzite structure with a preferred c-axis orientation regardless of the substrate. The surface morphology and refractive index were compared between the three samples and the mechanisms behind it were discussed.

Mukhopadhyay et al. (contribution 8) reported a crack-free AlGaN/AlN/GaN HEMT structure with a high aluminum composition (>35%) and thick barrier (>30 nm) grown on sapphire substrate. It exhibited ultra-low sheet resistivity (<250 $\Omega/\square$). The optimized growth conditions were detailed. The density of 2DEG was as high as 1.46×10^{13} cm^{-2} and the mobility reached 1710 cm^2/V·s at room temperature.

The ninth article (contribution 9) also comes from Mukhopadhyay et al., who prepared carbon-doped semi-insulating N-polar GaN on a sapphire substrate through a propane precursor. As N-polar GaN usually contains more oxygen than its Ga-polar counterparts, thus resulting in a high unintentionally doped electron concentration, this work provides a feasible method to grow semi-insulating N-polar GaN.

In Zhang et al.'s article (contribution 10), InGaN-based red microLEDs (μLEDs) of different sizes were prepared. The KOH wet treatment was investigated to alleviate the surface damage to sidewalls after dry etching. It could significantly inhibit the surface non-radiative recombination processes, thus enhancing the optical and electrical performances of the 5 μm μLEDs.

Han et al. (contribution 10) reviewed the research progress and development prospects of enhanced GaN HEMTs. The importance and merits of Si-based GaN HEMTs were illustrated. The MOCVD growth technology, HEMT structures, reliability, and CMOS compatibility were delineated and compared. Future development directions were also envisioned.

The twelfth text is also a review by Jafar et al. (contribution 12) on the growth conditions and EQEs of InGaN LEDs. The challenges of InGaN growth were reviewed. The mechanisms behind the efficiency droop were analyzed. Furthermore, novel approaches to improve the EQE were also discussed.

3. Conclusions

This compilation of articles is devoted to the growth, device fabrication, and theoretical calculations of III-nitride materials and structures. DUV LEDs, green lasers, red microLEDs, HEMTs, the low-temperature growth of III-nitride by PEALD, N-polar GaN, the heterostructure of GaN/MoS$_2$, vertical GaN Schottky diodes, and the mechanism of carrier transportation in multiple quantum wells are investigated by researchers globally. All the studies are important issues in the realm of III-nitride materials and applications. We hope the methods and results in this Special Issue will promote the exploration of III-nitride materials and applications, which keep providing discoveries that will change our daily lives.

Conflicts of Interest: The author declares no conflicts of interest.

List of Contributions

1. Li, Z.; Lu, H.; Wang, J.; Zhu, Y.; Yu, T.; Tian, Y. Maximizing the Light Extraction Efficiency for AlGaN-Based DUV-LEDs with Two Optimally Designed Surface Structures under the Guidance of PSO. *Crystals* **2022**, *12*, 1700. https://doi.org/10.3390/cryst12121700.
2. Chen, Z.; Liang, F.; Zhao, D.; Yang, J.; Chen, P.; Jiang, D. Investigation into the MOCVD Growth and Optical Properties of InGaN/GaN Quantum Wells by Modulating NH3 Flux. *Crystals* **2023**, *13*, 127. https://doi.org/10.3390/cryst13010127.
3. Tian, F.; Kong, D.; Qiu, P.; Liu, H.; Zhu, X.; Wei, H.; Song, Y.; Chen, H.; Zheng, X.; Peng, M. Polarization Modulation on Charge Transfer and Band Structures of GaN/MoS$_2$ Polar Heterojunctions. *Crystals* **2023**, *13*, 563. https://doi.org/10.3390/cryst13040563.
4. N'Dohi, A.J.E.; Sonneville, C.; Saidi, S.; Ngo, T.H.; De Mierry, P.; Frayssinet, E.; Cordier, Y.; Phung, L.V.; Morancho, F.; Maher, H.; et al. Micro-Raman Spectroscopy Study of Vertical GaN Schottky Diode. *Crystals* **2023**, *13*, 713. https://doi.org/10.3390/cryst13050713.
5. Zhang, Y.; Ding, G.; Wang, F.; Yu, P.; Feng, Q.; Yu, C.; He, J.; Wang, X.; Xu, W.; He, M.; et al. Normally-Off p-GaN Gate High-Electron-Mobility Transistors with the Air-Bridge Source-Connection Fabricated Using the Direct Laser Writing Grayscale Photolithography Technology. *Crystals* **2023**, *13*, 815. https://doi.org/10.3390/cryst13050815.
6. Guo, J.; Liu, W.; Ding, D.; Tan, X.; Zhang, W.; Han, L.; Wang, Z.; Gong, W.; Li, J.; Zhai, R.; et al. Analysis of Photo-Generated Carrier Escape in Multiple Quantum Wells. *Crystals* **2023**, *13*, 834. https://doi.org/10.3390/cryst13050834.
7. Liu, S.; Li, Y.; Tao, J.; Tang, R.; Zheng, X. Structural, Surface, and Optical Properties of AlN Thin Films Grown on Different Substrates by PEALD. *Crystals* **2023**, *13*, 910. https://doi.org/10.3390/cryst13060910.
8. Mukhopadhyay, S.; Liu, C.; Chen, J.; Tahmidul Alam, M.; Sanyal, S.; Bai, R.; Wang, G.; Gupta, C.; Pasayat, S.S. Crack-Free High-Composition (>35%) Thick-Barrier (>30 nm) AlGaN/AlN/GaN High-Electron-Mobility Transistor on Sapphire with Low Sheet

Resistance (<250 Ω/□). *Crystals* **2023**, *13*, 1456. https://doi.org/10.3390/cryst13101456.
9. Mukhopadhyay, S.; Sanyal, S.; Wang, G.; Gupta, C.; Pasayat, S.S. First Demonstration of Extrinsic C-Doped Semi-Insulating N-Polar GaN Using Propane Precursor Grown on Miscut Sapphire Substrate by MOCVD. *Crystals* **2023**, *13*, 1457. https://doi.org/10.3390/cryst13101457.
10. Zhang, S.; Fan, Q.; Ni, X.; Tao, L.; Gu, X. Study on the Influence of KOH Wet Treatment on Red μLEDs. *Crystals* **2023**, *13*, 1611. https://doi.org/10.3390/cryst13121611.
11. Han, L.; Tang, X.; Wang, Z.; Gong, W.; Zhai, R.; Jia, Z.; Zhang, W. Research Progress and Development Prospects of Enhanced GaN HEMTs. *Crystals* **2023**, *13*, 911. https://doi.org/10.3390/cryst13060911.
12. Jafar, N.; Jiang, J.; Lu, H.; Qasim, M.; Zhang, H. Recent Research on Indium-Gallium-Nitride-Based Light-Emitting Diodes: Growth Conditions and External Quantum Efficiency. *Crystals* **2023**, *13*, 1623. https://doi.org/10.3390/cryst13121623.

References

1. Amano, H.; Kito, M.; Hiramatsu, K.; Akasaki, I. P-Type Conduction in Mg-Doped GaN Treated with Low-Energy Electron Beam Irradiation (LEEBI). *Jpn. J. Appl. Phys.* **1989**, *28*, L2112–L2114. [CrossRef]
2. Nakamura, S.; Mukai, T.; Senoh, M.; Iwasa, N. Thermal Annealing Effects on P-Type Mg-Doped GaN Films. *Jpn. J. Appl. Phys.* **1992**, *31*, L139–L142. [CrossRef]
3. Nakamura, S.; Mukai, T.; Senoh, M. Candela-class high-brightness InGaN/AlGaN double-heterostructure blue-light-emitting diodes. *Appl. Phys. Lett.* **1994**, *64*, 1687–1689. [CrossRef]
4. Wang, J.; Xie, N.; Xu, F.; Zhang, L.; Lang, J.; Kang, X.; Qin, Z.; Yang, X.; Tang, N.; Wang, X.; et al. Group-III nitride heteroepitaxial films approaching bulk-class quality. *Nat. Mater.* **2023**, *22*, 853–859. [CrossRef] [PubMed]
5. Shen, J.; Yang, X.; Liu, D.; Cai, Z.; Wei, L.; Xie, N.; Xu, F.; Tang, N.; Wang, X.; Ge, W.; et al. High quality AlN film grown on a nano-concave-circle patterned Si substrate with an AlN seed layer. *Appl. Phys. Lett.* **2020**, *117*, 022103. [CrossRef]
6. Wang, J.; Xu, F.; Liu, B.; Lang, J.; Zhang, N.; Kang, X.; Qin, Z.; Yang, X.; Wang, X.; Ge, W.; et al. Control of dislocations in heteroepitaxial AlN films by extrinsic supersaturated vacancies introduced through thermal desorption of heteroatoms. *Appl. Phys. Lett.* **2021**, *118*, 162103. [CrossRef]
7. Wu, S.; Yang, X.; Zhang, H.; Shi, L.; Zhang, Q.; Shang, Q.; Qi, Z.; Xu, Y.; Zhang, J.; Tang, N.; et al. Unambiguous Identification of Carbon Location on the N Site in Semi-insulating GaN. *Phys. Rev. Lett.* **2018**, *121*, 145505. [CrossRef] [PubMed]
8. Li, Y.; Yan, S.; Hu, X.; Song, Y.; Deng, Z.; Du, C.; Wang, W.; Ma, Z.; Wang, L.; Jia, H.; et al. Characterization of edge dislocation density through X-ray diffraction rocking curves. *J. Cryst. Growth* **2020**, *551*, 125893. [CrossRef]
9. Li, Y.; Die, J.; Yan, S.; Deng, Z.; Ma, Z.; Wang, L.; Jia, H.; Wang, W.; Jiang, Y.; Chen, H. Characterization of periodicity fluctuations in InGaN/GaN MQWs by the kinematical simulation of X-ray diffraction. *Appl. Phys. Express* **2019**, *12*, 045502. [CrossRef]
10. Zhu, Y.; Lu, T.; Zhou, X.; Zhao, G.; Dong, H.; Jia, Z.; Liu, X.; Xu, B. Origin of huge photoluminescence efficiency improvement in InGaN/GaN multiple quantum wells with low-temperature GaN cap layer grown in N_2/H_2 mixture gas. *Appl. Phys. Express* **2017**, *10*, 061004. [CrossRef]
11. Li, Y.; Deng, Z.; Ma, Z.; Wang, L.; Jia, H.; Wang, W.; Jiang, Y.; Chen, H. Visualizing carrier transitions between localization states in a InGaN yellow–green light-emitting-diode structure. *J. Appl. Phys.* **2019**, *126*, 095705. [CrossRef]
12. Li, Y.; Li, Y.; Zhang, J.; Wang, Y.; Li, T.; Jiang, Y.; Jia, H.; Wang, W.; Yang, R.; Chen, H. Direct Observation of Carrier Transportation between Localized States in InGaN Quantum Wells. *Crystals* **2022**, *12*, 1837. [CrossRef]
13. Ma, J.; Ji, X.; Wang, G.; Wei, X.; Lu, H.; Yi, X.; Duan, R.; Wang, J.; Zeng, Y.; Li, J.; et al. Anomalous temperature dependence of photoluminescence in self-assembled InGaN quantum dots. *Appl. Phys. Lett.* **2012**, *101*, 131101. [CrossRef]
14. Lu, T.; Ma, Z.; Du, C.; Fang, Y.; Wu, H.; Jiang, Y.; Wang, L.; Dai, L.; Jia, H.; Liu, W.; et al. Temperature-dependent photoluminescence in light-emitting diodes. *Sci. Rep.* **2014**, *4*, 6131. [CrossRef] [PubMed]
15. Liu, W.; Zhao, D.G.; Jiang, D.S.; Chen, P.; Liu, Z.S.; Zhu, J.J.; Shi, M.; Zhao, D.M.; Li, X.; Liu, J.P.; et al. Localization effect in green light emitting InGaN/GaN multiple quantum wells with varying well thickness. *J. Alloys Compd.* **2015**, *625*, 266–270. [CrossRef]
16. Weng, G.-E.; Zhao, W.-R.; Chen, S.-Q.; Akiyama, H.; Li, Z.-C.; Liu, J.-P.; Zhang, B.-P. Strong localization effect and carrier relaxation dynamics in self-assembled InGaN quantum dots emitting in the green. *Nanoscale Res. Lett.* **2015**, *10*, 31. [CrossRef] [PubMed]
17. Yang, J.; Zhao, D.G.; Jiang, D.S.; Chen, P.; Zhu, J.J.; Liu, Z.S.; Le, L.C.; Li, X.J.; He, X.G.; Liu, J.P.; et al. Optical and structural characteristics of high indium content InGaN/GaN multi-quantum wells with varying GaN cap layer thickness. *J. Appl. Phys.* **2015**, *117*. [CrossRef]
18. Li, Y.; Jin, Z.; Han, Y.; Zhao, C.; Huang, J.; Tang, C.W.; Wang, J.; Lau, K.M. Surface morphology and optical properties of InGaN quantum dots with varying growth interruption time. *Mater. Res. Express* **2019**, *7*, 015903. [CrossRef]
19. Li, Y.; Yang, R.; Jiang, Y.; Jia, H.; Wang, W.; Chen, H. In Situ AlGaN Interlayer for Reducing the Reverse Leakage Current of InGaN Light-Emitting Diodes. *IEEE Electron. Device Lett.* **2023**, *44*, 777–780. [CrossRef]

20. Jiang, Y.; Li, Y.; Li, Y.; Deng, Z.; Lu, T.; Ma, Z.; Zuo, P.; Dai, L.; Wang, L.; Jia, H.; et al. Realization of high-luminous-efficiency InGaN light-emitting diodes in the "green gap" range. *Sci. Rep.* **2015**, *5*, 10883. [CrossRef]
21. Bi, Z.; Lenrick, F.; Colvin, J.; Gustafsson, A.; Hultin, O.; Nowzari, A.; Lu, T.; Wallenberg, R.; Timm, R.; Mikkelsen, A.; et al. InGaN Platelets: Synthesis and Applications toward Green and Red Light-Emitting Diodes. *Nano Lett.* **2019**, *19*, 2832–2839. [CrossRef]
22. Iida, D.; Zhuang, Z.; Kirilenko, P.; Velazquez-Rizo, M.; Najmi, M.A.; Ohkawa, K. 633-nm InGaN-based red LEDs grown on thick underlying GaN layers with reduced in-plane residual stress. *Appl. Phys. Lett.* **2020**, *116*, 162101. [CrossRef]
23. Kneissl, M.; Seong, T.-Y.; Han, J.; Amano, H. The emergence and prospects of deep-ultraviolet light-emitting diode technologies. *Nat. Photonics* **2019**, *13*, 233–244. [CrossRef]
24. Lee, D.-g.; Choi, Y.; Jung, S.; Kim, Y.; Park, S.; Choi, P.; Yoon, S. High-efficiency InGaN red light-emitting diodes with external quantum efficiency of 10.5% using extended quantum well structure with AlGaN interlayers. *Appl. Phys. Lett.* **2024**, *124*, 121109. [CrossRef]
25. Liu, C.; Ooi, Y.K.; Islam, S.M.; Xing, H.G.; Jena, D.; Zhang, J. 234 nm and 246 nm AlN-Delta-GaN quantum well deep ultraviolet light-emitting diodes. *Appl. Phys. Lett.* **2018**, *112*, 011101. [CrossRef]
26. Xing, K.; Hu, J.; Pan, Z.; Xia, Z.; Jin, Z.; Wang, L.; Jiang, X.; Wang, H.; Zeng, H.; Wang, X. Demonstration of 651 nm InGaN-based red light-emitting diode with an external quantum efficiency over 6% by InGaN/AlN strain release interlayer. *Opt. Express* **2024**, *32*, 11377. [CrossRef]
27. Zhang, S.; Zhang, J.; Gao, J.; Wang, X.; Zheng, C.; Zhang, M.; Wu, X.; Xu, L.; Ding, J.; Quan, Z.; et al. Efficient emission of InGaN-based light-emitting diodes: Toward orange and red. *Photonics Res.* **2020**, *8*, 1671. [CrossRef]
28. Narukawa, Y.; Ichikawa, M.; Sanga, D.; Sano, M.; Mukai, T. White light emitting diodes with super-high luminous efficacy. *J. Phys. D-Appl. Phys.* **2010**, *43*, 354002. [CrossRef]
29. Lv, Q.; Liu, J.; Mo, C.; Zhang, J.; Wu, X.; Wu, Q.; Jiang, F. Realization of Highly Efficient InGaN Green LEDs with Sandwich-like Multiple Quantum Well Structure: Role of Enhanced Interwell Carrier Transport. *ACS Photonics* **2018**, *6*, 130–138. [CrossRef]
30. Chen, Z.; Yan, S.; Danesh, C. MicroLED technologies and applications: Characteristics, fabrication, progress, and challenges. *J. Phys. D-Appl. Phys.* **2021**, *54*, 123001. [CrossRef]
31. Jiang, F.; Zhang, J.; Xu, L.; Ding, J.; Wang, G.; Wu, X.; Wang, X.; Mo, C.; Quan, Z.; Guo, X.; et al. Efficient InGaN-based yellow-light-emitting diodes. *Photonics Res.* **2019**, *7*, 144. [CrossRef]
32. Zhou, X.; Lu, T.; Zhu, Y.; Zhao, G.; Dong, H.; Jia, Z.; Yang, Y.; Chen, Y.; Xu, B. Surface Morphology Evolution Mechanisms of InGaN/GaN Multiple Quantum Wells with Mixture N_2/H_2-Grown GaN Barrier. *Nanoscale Res. Lett.* **2017**, *12*, 354. [CrossRef] [PubMed]
33. Wu, Q.-f.; Cao, S.; Mo, C.-l.; Zhang, J.-l.; Wang, X.-l.; Quan, Z.-j.; Zheng, C.-d.; Wu, X.-m.; Pan, S.; Wang, G.-X.; et al. Effects of Hydrogen Treatment in Barrier on the Electroluminescence of Green InGaN/GaN Single-Quantum-Well Light-Emitting Diodes with V-Shaped Pits Grown on Si Substrates. *Chin. Phys. Lett.* **2018**, *35*, 098501. [CrossRef]
34. Li, Y.; Yan, S.; Hu, X.; Song, Y.; Deng, Z.; Du, C.; Wang, W.; Ma, Z.; Wang, L.; Jia, H.; et al. Effect of H_2 treatment in barrier on interface, optical and electrical properties of InGaN light emitting diodes. *Superlattices Microstruct.* **2020**, *145*, 106606. [CrossRef]
35. Li, Y.; Yan, S.; Die, J.; Hu, X.; Song, Y.; Deng, Z.; Du, C.; Wang, W.; Ma, Z.; Wang, L.; et al. The influence of excessive H2 during barrier growth on InGaN light-emitting diodes. *Mater. Res. Express* **2020**, *7*, 105907. [CrossRef]
36. Li, Y.; Liu, C.; Zhang, Y.; Jiang, Y.; Hu, X.; Song, Y.; Su, Z.; Jia, H.; Wang, W.; Chen, H. Realizing Single Chip White Light InGaN LED via Dual-Wavelength Multiple Quantum Wells. *Materials* **2022**, *15*, 3998. [CrossRef] [PubMed]
37. Behrman, K.; Kymissis, I. Micro light-emitting diodes. *Nat. Electron.* **2022**, *5*, 564–573. [CrossRef]
38. Sheen, M.; Ko, Y.; Kim, D.-u.; Kim, J.; Byun, J.-h.; Choi, Y.; Ha, J.; Yeon, K.Y.; Kim, D.; Jung, J.; et al. Highly efficient blue InGaN nanoscale light-emitting diodes. *Nature* **2022**, *608*, 56–61. [CrossRef]
39. Wang, L.; Ma, J.; Su, P.; Huang, J. High-Resolution Pixel LED Headlamps: Functional Requirement Analysis and Research Progress. *Appl. Sci.* **2021**, *11*, 3368. [CrossRef]
40. Xiong, J.; Hsiang, E.-L.; He, Z.; Zhan, T.; Wu, S.-T. Augmented reality and virtual reality displays: Emerging technologies and future perspectives. *Light-Sci. Appl.* **2021**, *10*, 216. [CrossRef]
41. Yu, L.; Wang, L.; Yang, P.; Hao, Z.; Yu, J.; Luo, Y.; Sun, C.; Xiong, B.; Han, Y.; Wang, J.; et al. Metal organic vapor phase epitaxy of high-indium-composition InGaN quantum dots towards red micro-LEDs. *Opt. Mater. Express* **2022**, *12*, 3225. [CrossRef]
42. Zhuang, Z.; Iida, D.; Ohkawa, K. Ultrasmall and ultradense InGaN-based RGB monochromatic micro-light-emitting diode arrays by pixilation of conductive p-GaN. *Photonics Res.* **2021**, *9*, 2429. [CrossRef]
43. Chen, Z.; Sheng, B.; Liu, F.; Liu, S.; Li, D.; Yuan, Z.; Wang, T.; Rong, X.; Huang, J.; Qiu, J.; et al. High-Efficiency InGaN Red Mini-LEDs on Sapphire Toward Full-Color Nitride Displays: Effect of Strain Modulation. *Adv. Funct. Mater.* **2023**, *33*, 2300042. [CrossRef]
44. Li, P.; Li, H.; Yao, Y.; Lim, N.; Wong, M.; Iza, M.; Gordon, M.J.; Speck, J.S.; Nakamura, S.; DenBaars, S.P. Significant Quantum Efficiency Enhancement of InGaN Red Micro-Light-Emitting Diodes with a Peak External Quantum Efficiency of up to 6%. *ACS Photonics* **2023**, *10*, 1899–1905. [CrossRef]
45. Sun, Y.; Zhou, K.; Sun, Q.; Liu, J.; Feng, M.; Li, Z.; Zhou, Y.; Zhang, L.; Li, D.; Zhang, S.; et al. Room-temperature continuous-wave electrically injected InGaN-based laser directly grown on Si. *Nat. Photonics* **2016**, *10*, 595–599. [CrossRef]
46. Yang, T.; Chen, Y.-H.; Wang, Y.-C.; Ou, W.; Ying, L.-Y.; Mei, Y.; Tian, A.-Q.; Liu, J.-P.; Guo, H.-C.; Zhang, B.-P. Green Vertical-Cavity Surface-Emitting Lasers Based on InGaN Quantum Dots and Short Cavity. *Nano-Micro Lett.* **2023**, *15*, 223. [CrossRef] [PubMed]

47. Raun, A.; Hu, E. Ultralow Thresh. Blue Quantum Dot Lasers: What's True Recipe Success? *Nanophotonics* **2020**, *10*, 23–29. [CrossRef]
48. Wang, D.; Zhu, T.; Oliver, R.A.; Hu, E.L. Ultra-low-threshold InGaN/GaN quantum dot micro-ring lasers. *Opt. Lett.* **2018**, *43*, 799–802. [CrossRef] [PubMed]
49. Feng, M.; Zhao, H.; Zhou, R.; Tang, Y.; Liu, J.; Sun, X.; Sun, Q.; Yang, H. Continuous-Wave Current Injected InGaN/GaN Microdisk Laser on Si(100). *ACS Photonics* **2022**, *10*, 2208–2215. [CrossRef]
50. Zi, H.; Fu, W.Y.; Cheung, Y.F.; Damilano, B.; Frayssinet, E.; Alloing, B.; Duboz, J.-Y.; Boucaud, P.; Semond, F.; Choi, H.W. Comparison of lasing characteristics of GaN microdisks with different structures. *J. Phys. D-Appl. Phys.* **2022**, *55*, 355107. [CrossRef]
51. Tajiri, T.; Sosumi, S.; Shimoyoshi, K.; Uchida, K. Fabrication and optical characterization of GaN micro-disk cavities undercut by laser-assisted photo-electrochemical etching. *Jpn. J. Appl. Phys.* **2023**, *62*, SC1069. [CrossRef]
52. Zhao, L.; Chen, J.; Liu, C.; Lin, S.; Ge, X.; Li, X.; Hu, T.; Ding, S.; Wang, K. Low-threshold InGaN-based whispering gallery mode laser with lateral nanoporous distributed Bragg reflector. *Opt. Laser Technol.* **2023**, *164*, 109480. [CrossRef]
53. Chen, K.J.; Häberlen, O.; Lidow, A.; Tsai, C.I.; Ueda, T.; Uemoto, Y.; Wu, Y. GaN-on-Si Power Technology: Devices and Applications. *IEEE Trans. Electron. Devices* **2017**, *64*, 779–795. [CrossRef]
54. Cai, Y.; Zhou, Y.; Chen, K.J.; Lau, K.M. High-performance enhancement-mode AlGaN/GaN HEMTs using fluoride-based plasma treatment. *IEEE Electron. Device Lett.* **2005**, *26*, 435–437. [CrossRef]
55. Cui, J.; Wei, J.; Wang, M.; Wu, Y.; Yang, J.; Li, T.; Yu, J.; Yang, H.; Yang, X.; Wang, J.; et al. 6500-V E-mode Active-Passivation p-GaN Gate HEMT with Ultralow Dynamic RON. In Proceedings of the 2023 International Electron Devices Meeting (IEDM), San Francisco, CA, USA, 9–13 December 2023; pp. 1–4. [CrossRef]
56. Zheng, Z.; Song, W.; Zhang, L.; Yang, S.; Wei, J.; Chen, K.J. High I_{ON} and I_{ON}/I_{OFF} Ratio Enhancement-Mode Buried p-Channel GaN MOSFETs on p-GaN Gate Power HEMT Platform. *IEEE Electron. Device Lett.* **2020**, *41*, 26–29. [CrossRef]
57. Zheng, Z.; Zhang, L.; Song, W.; Feng, S.; Xu, H.; Sun, J.; Yang, S.; Chen, T.; Wei, J.; Chen, K.J. Gallium nitride-based complementary logic integrated circuits. *Nat. Electron.* **2021**, *4*, 595–603. [CrossRef]
58. Chen, J.; Liu, Z.; Wang, H.; He, Y.; Zhu, X.; Ning, J.; Zhang, J.; Hao, Y. A GaN Complementary FET Inverter With Excellent Noise Margins Monolithically Integrated With Power Gate-Injection HEMTs. *IEEE Trans. Electron. Devices* **2022**, *69*, 51–56. [CrossRef]
59. Xie, Q.; Yuan, M.; Niroula, J.; Sikder, B.; Greer, J.A.; Rajput, N.S.; Chowdhury, N.; Palacios, T. Highly Scaled GaN Complementary Technology on a Silicon Substrate. *IEEE Trans. Electron. Devices* **2023**, *70*, 2121–2128. [CrossRef]
60. Keller, S.; Li, H.; Laurent, M.; Hu, Y.; Pfaff, N.; Lu, J.; Brown, D.F.; Fichtenbaum, N.A.; Speck, J.S.; DenBaars, S.P.; et al. Recent progress in metal-organic chemical vapor deposition of (000) N-polar group-III nitrides. *Semicond. Sci. Technol.* **2014**, *29*, 113001. [CrossRef]
61. Li, Y.; Jiang, Y.; Jia, H.; Wang, W.; Yang, R.; Chen, H. Superior Optoelectronic Performance of N-Polar GaN LED to Ga-Polar Counterpart in the "Green Gap" Range. *IEEE Access* **2022**, *10*, 95565–95570. [CrossRef]
62. Li, C.; Zhang, K.; Qiaoyu, Z.; Yin, X.; Ge, X.; Wang, J.; Wang, Q.; He, C.; Zhao, W.; Chen, Z. High quality N-polar GaN films grown with varied V/III ratios by metal–organic vapor phase epitaxy. *RSC Adv.* **2020**, *10*, 43187–43192. [CrossRef] [PubMed]
63. Li, Y.; Hu, X.; Song, Y.; Su, Z.; Wang, W.; Jia, H.; Wang, W.; Jiang, Y.; Chen, H. Epitaxy N-polar GaN on vicinal Sapphire substrate by MOCVD. *Vacuum* **2021**, *189*, 110173. [CrossRef]
64. Li, Y.; Hu, X.; Song, Y.; Su, Z.; Wang, W.; Jia, H.; Wang, W.; Jiang, Y.; Chen, H. The role of AlN thickness in MOCVD growth of N-polar GaN. *J. Alloys Compd.* **2021**, *884*, 161134. [CrossRef]
65. Li, Y.; Hu, X.; Song, Y.; Su, Z.; Jia, H.; Wang, W.; Jiang, Y.; Chen, H. The influence of temperature of nitridation and AlN buffer layer on N-polar GaN. *Mater. Sci. Semicond. Process.* **2022**, *141*, 106423. [CrossRef]
66. Yamada, S.; Shirai, M.; Kobayashi, H.; Arai, M.; Kachi, T.; Suda, J. Realization of low specific-contact-resistance on N-polar GaN surfaces using heavily-Ge-doped n-type GaN films deposited by low-temperature reactive sputtering technique. *Appl. Phys. Express* **2024**, *17*, 036501. [CrossRef]
67. Romanczyk, B.; Wienecke, S.; Guidry, M.; Li, H.; Ahmadi, E.; Zheng, X.; Keller, S.; Mishra, U.K. Demonstration of Constant 8 W/mm Power Density at 10, 30, and 94 GHz in State-of-the-Art Millimeter-Wave N-Polar GaN MISHEMTs. *IEEE Trans. Electron. Devices* **2018**, *65*, 45–50. [CrossRef]
68. Koksaldi, O.S.; Haller, J.; Li, H.; Romanczyk, B.; Guidry, M.; Wienecke, S.; Keller, S.; Mishra, U.K. N-Polar GaN HEMTs Exhibiting Record Breakdown Voltage Over 2000 V and Low Dynamic On-Resistance. *IEEE Electron Device Lett.* **2018**, *39*, 1014–1017. [CrossRef]
69. Hamwey, R.; Hatui, N.; Akso, E.; Wu, F.; Clymore, C.; Keller, S.; Speck, J.S.; Mishra, U.K. First Demonstration of an N-Polar InAlGaN/GaN HEMT. *IEEE Electron Device Lett.* **2024**, *45*, 328–331. [CrossRef]

Disclaimer/Publisher's Note: The statements, opinions and data contained in all publications are solely those of the individual author(s) and contributor(s) and not of MDPI and/or the editor(s). MDPI and/or the editor(s) disclaim responsibility for any injury to people or property resulting from any ideas, methods, instructions or products referred to in the content.

Article

Maximizing the Light Extraction Efficiency for AlGaN-Based DUV-LEDs with Two Optimally Designed Surface Structures under the Guidance of PSO

Zizheng Li [1], Huimin Lu [1,*], Jianping Wang [1], Yifan Zhu [1], Tongjun Yu [2] and Yucheng Tian [2]

1. School of Computer and Communication Engineering, University of Science and Technology Beijing, Beijing 100083, China
2. The State Key Laboratory for Mesoscopic Physics, School of Physics, Peking University, Beijing 100871, China
* Correspondence: hmlu@ustb.edu.cn

Abstract: A novel method of utilizing an intelligent algorithm to guide the light extraction surface structure designing process for deep-ultraviolet light emitting diodes (DUV-LEDs) is proposed and investigated. Two kinds of surface structures based on the truncated pyramid array (TPA) and truncated cone array (TCA) are applied, which are expected to suppress the total internal reflection (TIR) effect and increase the light extraction efficiency (LEE). By addressing particle swarm optimization (PSO), the TPA and TCA microstructures constructed on the sapphire layer of the flip-chip DUV-LEDs are optimized. Compared to the conventional structure design method of parameter sweeping, this algorithm has much higher design efficiency and better optical properties. At the DUV wavelength of 280 nm, as a result, significant increases of 221% and 257% on the LEE are realized over the two forms of optimized surface structures. This approach provides another design path for DUV-LED light extraction structures.

Keywords: DUV-LED; light extraction efficiency; particle swarm algorithm; microstructure optimization

1. Introduction

AlGaN-based deep-ultraviolet light-emitting diodes (DUV-LEDs) have widely drawn attention from academia and industry for their promising potential in sanitation disinfection, medical diagnosis, microstructure lithography, confidential communication, the fight against coronavirus, and so on [1–4]. In spite of the numerous efforts by predecessors [5], the external quantum efficiency (EQE) of DUV-LEDs still remains extremely low, resulting in a much weaker output power compared with the common LEDs in other wavelength ranges. Two major causes of this are the high-density dislocation in epitaxial layers and the extremely low light extraction efficiency (LEE) [6,7]. Fabricating DUV-LEDs on the AlN substrate can overcome the dislocation problem, meanwhile enhancing the device's reliability. To avoid the high absorption of DUV light in the P-region, flip-chip configuration is widely used such that the light extraction process happens in the transparent sapphire layer. However, high refractive index contrast at the interface between the epitaxial layer and ambient medium induces total internal reflection (TIR) in a wide range of incidence angles, which can lead to a significant decline of the extracted light from the LEDs surface. It has been proved that at the wavelength of 280 nm, mostly TE [8]. Therefore, a range of studies on light extraction enhancement by using certain microstructures has been performed, including moth-eye structure array, cone array, nanoimprinted structures, and so on [9–13]. However, systematic optimization of such surface structures has not yet been reported.

Remarkable progress has been made by intelligent algorithms in the design process of photonic devices, for the structure optimization design is such an extremely time-consuming process that it requires enormous computing power if the conventional working flow, which relies on the human brain and experience along with exhaustive parameter sweeps, is adopted [14]. In general, the intelligent algorithm-driven design methods are firstly initialized by a rational precondition (obtained from human guess or experience) and iterate by calculating the performances then updating the parameters and photonic structure for multiple generations to satisfy the ultimate expectations [15–18]. Broadly speaking, intelligent algorithms that are addressed in device design include optimization algorithms (e.g., particle swarm optimization (PSO) and genetic algorithm (GA)), topological optimization (TO), and deep learning (DL) algorithms (e.g., generative adversarial network (GAN)). The design process of the three-dimensional devices addressing has been illustrated. For the task of designing a light extraction structure for DUV-LEDs that demands a huge amount of calculation time in one single simulation, along with multiple parameters linearly and non-linearly affecting the optical response that need to be determined, PSO is the most effective and accurate method in comparison with other intelligent algorithms. The traditional and most applied method, parameter sweep, searches every grid point in the grid net to find an optimal combination of parameters, which is certainly inefficient and not worthy when too many simulation runs are required. DL requires a large-scale dataset and is greatly dependent on the quality of the dataset. Besides, it could become inoperable to build such a dataset if each sample is obtained by a time-consuming FDTD simulation method. Apart from that, methods such as GA are low efficient for the slow convergence with the mutation and exchange procedure. Both of them provide strong resistance to being trapped in local optimum, which overkill the tasks with some number of parameters to determine, such as the one here for DUV-LEDs.

In this work, we propose a novel design scheme for the light extraction surface structures of AlGaN-based DUV-LEDs. By modifying the structural parameters under the guidance of an intelligent algorithm, we maximize the LEE. On this basis, two kinds of light extraction surface microstructures, a truncated pyramid array and a truncated cone array, are applied and optimized. The basic theory and properties of the microstructures are discussed first. Then, the parameter optimization trends and the changes in optical responses during the algorithm iterations are identified using three-dimensional (3D) finite difference time domain (FDTD) simulations. Finally, compared with the conventional flat sapphire surface, the superior optical properties contributed from the optimized surface structures using the proposed design scheme are analyzed and discussed for DUV-LEDs.

2. Methods

The perspective view of the two kinds of adopted surface structures is given in Figure 1, which is expected to improve the DUV-LEDs LEE through theoretical prediction. As shown in Figure 1, in order to minimize TIR, two light extraction surface structures, truncated pyramid array (TPA) and truncated cone array (TCA), are applied to be planted at the interface of sapphire and air. The thickness of the sapphire base (substrate under TPA and TCA), AlN layer, and the n-AlGaN layer are 2 µm, 1 µm, and 2 µm, respectively. Below them, there are multiple-quantum-well (MQW) layers, p-AlGaN layers, and p-GaN layers (shown in the subgraph), all of which the thickness is 0.1 µm. The side lengths of the top and bottom sides of the truncated pyramid are d_1 and d_2, while the radii of that in truncated cone are marked as r_1 and r_2, respectively. Parameter hp and hc represent the height of the TPA and TCA. The refractive indexes of the sapphire, AlN, n-AlGaN, MQW, p-AlGaN, and p-GaN are 1.823, 2.16, 2.6, 2.7, 2.6, and 2.9, respectively. Additionally, the absorption coefficients of the AlGaN, MQW, and GaN are set to 170,000 cm^{-1}, 1000 cm^{-1}, and 10 cm^{-1} [10], respectively. It is assumed that the absorption induced by sapphire and AlN layers can be ignored. At the sapphire-air interface, there is a huge refractive index change of 0.823, which can cause that any light injected with an incident angle bigger than 33° will be reflected back to the sapphire medium. We introduced the TPA and TCA surface structures to guide this part of light into the periodical microstructures, that way it can propagate into air directly or after a couple of times of reflection.

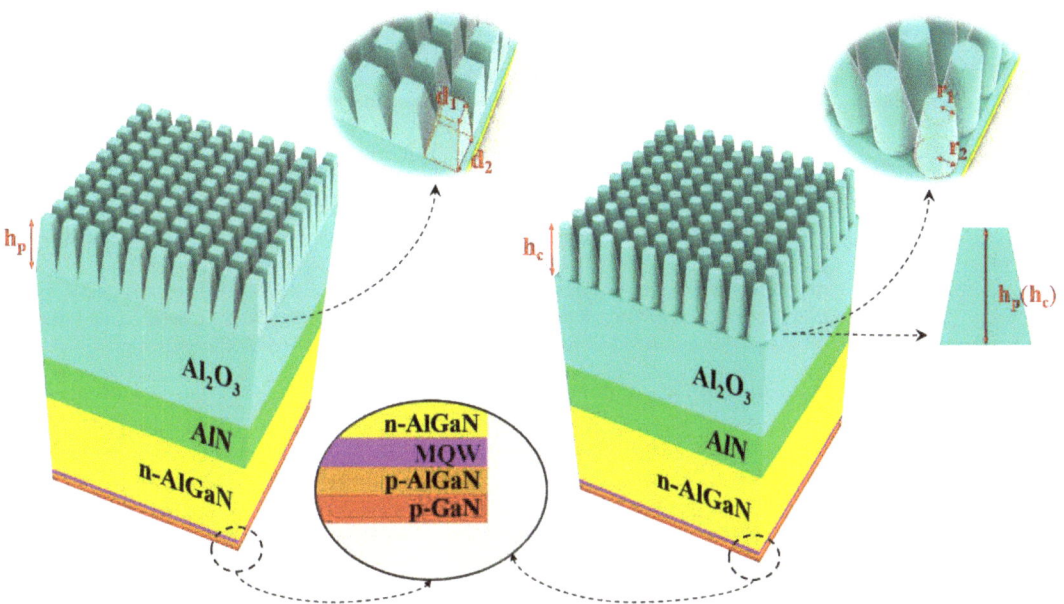

Figure 1. Perspective view of truncated-pyramid surface and truncated-cone surface DUV-LEDs.

In this work, the intelligent algorithm of PSO is utilized to search for the optimal parameter solution set of the surface microstructures, in order to gain the highest DUV-LEDs LEE. First, the microstructures are digitalized into three sets of structural parameters, height h, top scale d_1, r_1, bottom scale d_2, r_2. These parameters uniquely determine the structure of the TPA and TCA, given that the truncated pyramids and cones abut on each other. The parameter space is expressed by particles in PSO and the swarm consists of m particles as follows: S = {X_1, X_2, ... , X_m}, where each particle is an n-dimensional vector as follows: X_m = {$x_{m,1}$, $x_{m,2}$, ... , $x_{m,n}$}. Here, the number of particles is set to m = 6, which can lessen redundant computation and reduce time consumption. The particle dimension indicates the parameter space. Therefore, n = 3 and the vector for TPA is $x_{m,n}$ = {d_1, d_2, h} while for TPC is $x_{m,n}$ = {r_1, r_2, h}. Before running the PSO, the particles are randomly located in predetermined range and given random altering velocities [15]. Limited by the material growth in experiment, parameter h is preset within the range of [0 μm, 5 μm], and bottom scale parameters d_2, r_2 within the range of [0 μm, 1 μm]. The top scale parameters d_1, r_1 are set to values that are always not bigger than d_2 and r_2 respectively, ensuring keep in the shape of truncated pyramid and truncated cone. The figure of merits (FOM) is defined as the total power extracted from the LED at wavelength 280 nm. Therefore, the overall goal is to find the optimal solution that maximum the FOM. During the PSO iterative process, the particle location and velocity are updated by Equations (1) and (2) as follows:

$$x_{m,n} = x_{m,n} + v_{m,n} \qquad (1)$$

$$v_{m,n} = \omega \times v_{m,n} + c_1 \times rand \times (p_{best,mn} - x_{m,n}) + c_2 \times rand \times (g_{best,mn} - x_{m,n}) \qquad (2)$$

where $p_{best,mn}$ is the individual best position of the particle and $g_{best,mn}$ is the global best position of the swarm. Additionally, inertial weight ω determines the intention of the particle that prefers to keep the old velocity. The cognitive rate and social rate, c_1 and c_2, are to adjust the influence of individual and global best solutions. The random number *rand* is uniformly distributed in the range of [0, 1]. The PSO process ends when either of the following two terminating conditions are satisfied: 1. FOM shows no change in the last generations; 2. PSO hit the maximum generation number.

The 3D-FDTD algorithm is applied to perform the FOM calculation in each PSO iteration. A single dipole source, of which the wavelength and line width are 280 nm and 10 nm respectively, is placed at the center of the MQW region. Note that the TM mode is ignored in DUV-LEDs around the wavelength of 280 nm [8]. Therefore, the dipole source is polarized in the direction that is parallel to the MQW plane to only excite the TE mode. It is known that 3D-FDTD simulation meets great difficulty in large structures that exceed the scale of a hundred times of wavelength. To avoid too much time consuming, we limit the horizontal scale of the simulated region to approximately 4 μm × 4 μm. It is much smaller than the actual LED size, though still can reflect the real situation by covering the perfect electrical conductor (PEC) blocks on lateral sides of simulated region, which are acting as perfect mirrors that stop any power escaping laterally [19–21]. The extracted power is detected by the monitor at the top of the simulation region. The LEE is calculated in terms of the power extracted from the top side of the LED divided by the source power.

3. Results

In this work, two forms of light extraction microstructures, TPA and TCA, are analyzed and optimized by PSO at the sapphire-air interface, separately, in order to maximize the LEE of AlGaN-based DUV-LEDs. There are 6 particles in both tasks, marked as a, b, c, d, e, and f. Figure 2a,b gives the updated trends of the particles in the PSO progress for the two surface microstructures. It takes 12 generations of PSO iterations to find the optimal solution for the TPA structure, as shown in Figure 2a, while two more generations are taken for the TCA, shown in Figure 2b. These two sub-figures share the same color bar that indicates the specific LEE of a certain particle in a certain generation. In the first generation of PSO, the structural parameters are preset as the circumstance that there is no light extraction structure, and they are allocated with a random updating velocity. Particles then vary to find the individual best position in each generation and eventually gather at the global best position. The preset maximum generation numbers for the two tasks are both 20. However, neither of them hits it, and they both terminated by the terminating condition of no raise on FOM for the last three generations. It is worth to be noted that the preset conditions play a crucial part in the PSO procedure. We tested that if the number of particles is smaller than five, they may fall into a local maximum and find it difficult to jump out, and the PSO will not be terminated even after 20 generations. That leads to wrong solutions and time wasting. Nevertheless, drawing too many particles can reach the right solution but can also be time-consuming because the total time spent is the product of the number of particles, the number of generations, and the time of a single 3D-FDTD simulation.

In Figure 2c, the LEE varying trends over the PSO generations for the AlGaN-based DUV LED with the two surface microstructures are given and compared with that for the conventional LED. The yellow line labeled "Convention" represents a conventional LED without any light extraction structure. As for the TPA and TCA, in the first two generations, the search range is rapidly narrowed down, where the LEE has reached more than 10%. In the generations of 3~10, there are steadily minor increases in LEE. After generation 10, the changes in LEE become so small that eventually, the PSO is stopped by the first terminating condition. The values of $p_{best,mn}$ and $g_{best,mn}$ are updated during this procedure after each generation, and finally, the FOM reaches its maximum. It is worth mentioning that the LEEs obtained in the 13th and 14th generations are not included in Figure 2c because the LEE value in the last 3 generations remains the same. In comparison, the final optimal LEE of the LED with TPA is higher than that of the TCA because TPA can cover the entire sapphire-air interface more densely so that the light propagating through the sapphire layer will be fully guided into the microstructures. A 221% increment for TCA and a 257% increment for TPA are realized when compared to conventional LED.

The updating trends of the structural parameters for the two surface microstructures are also analyzed and given in Figure 3, which records the value of each parameter at the end of every generation in the PSO process. Figure 3a,b show the three parameters $\{d_1, d_2, h\}$, and it contains $\{r_1, r_2, h\}$ for TPA and TCA microstructure, respectively. Solid lines corresponding to the left y-axis are applied to the side length and diameter, while the dashed line corresponding to the right y-axis is applied to the height. As shown in Figure 3, the optimal solution obtained by the PSO is $\{d_1, d_2, h\} = \{0.31, 0.13, 5\}$, and $\{r_1, r_2, h\} = \{0.58, 0.19, 5\}$. In both cases for the TPA and TCA, the height h has been optimized to the preset range maximum value of 5 µm, suggesting that the height is a linear impact factor to the LEE. Actually, the higher the pyramids and cones are, the bigger the LEE. Additionally, if the microstructures can be extended upward infinitely, there will no longer be any TIR and the LEE encounters no decay. The parameters of side length and diameter show nonlinearity, and they fluctuate through the whole process with gradually decreasing amplitude. Finally, when the PSO is terminated, all the parameters tend to be stationary.

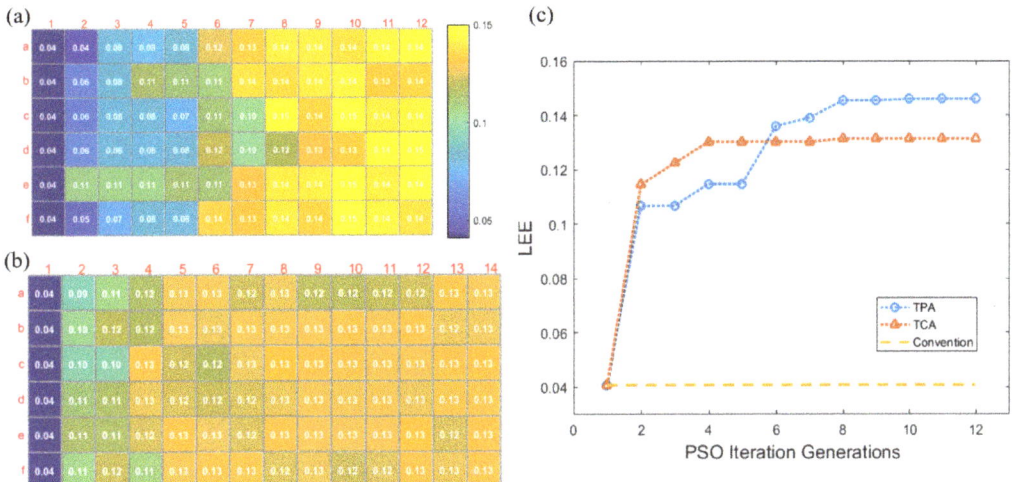

Figure 2. Updating trends of the particles in the structure of (**a**) the TPA and (**b**) the TCA, and (**c**) the comparison on the LEE enhancement trends.

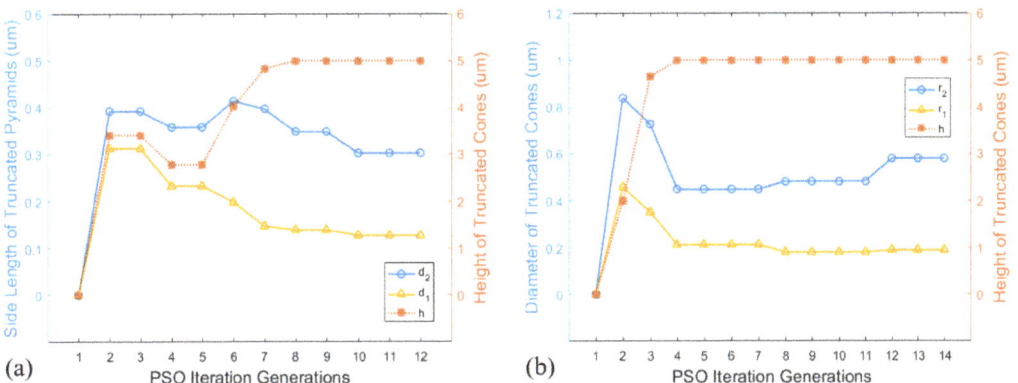

Figure 3. Updating trends of the structural parameters in (**a**) the TPA and (**b**) the TCA.

In this work, the task of determining structural parameters in TPA and TCA can be abstracted as finding the maximum of the function of multiple variables. The nonlinear relationship between the parameters and the LEE and the time-consuming FDTD calculations are the notable difficulty. The advantages of the PSO algorithm are highlighted in this procedure since there is no necessity to draw support from parameter sweeps or enumeration attempts. Suppose that we use the conventional method of parameter sweeping to determine such a structure with multiple parameters; even though they are independent of each other, we still have to split the whole task and spend a lot of time to find out how each parameter affects the LEE. This process deals with the parameters separately, it works only if the effect of the parameters on the FOM is linear, but it becomes hard to tell the way they affect the LEE jointly. In that case, all sorts of combinations of certain parameters need to be enumerated, and every time, a 3D-FDTD simulation is needed. However, with just a complete PSO procedure, the optimal values of all the parameters of a microstructure can be determined, greatly reducing the number of runs of FDTD simulations.

The optical field distribution for the AlGaN-based DUV-LEDs with the optimized surface microstructure is further investigated and compared with the conventional LEDs. The cross-sectional field distributions and light power variation trend in the direction horizontal to the MQW plane are shown in Figure 4. In the light extraction process, the biggest LEE decrease happens at the interface between the MQW layer and the n-AlGaN layer. More than half of the source power has been trapped in the MQW layer due to the high absorption and TIR at the interface. As depicted in Figure 4a, in the n-AlGaN layer and AlN layer, relatively low absorption occurs. Apart from the one near the dipole source, interfaces between every two adjoint layers cause TIR, which turns out to be the primary negative impact on light extraction. At the Al2O3-air interface, the normalized powers (which are actually the LEEs) influenced by three kinds of structures (no structure, TPA, and TCA) are compared. It can be seen that without any light extraction structures, the LEE of LED can finally be significantly low, even smaller than 5%. The TPA and TCA make it to raise the LEEs higher than 14.6% and 13.1%. In Figure 4b,c, the field distributions of the cross-section that is perpendicular to the MQW plane are illustrated. The white short, dashed lines are applied to highlight the boundaries of each layer. The power decrease in the light while propagating is shown in a more clear and visualized way. In the last part of the journey, where the effect of the TPA and TCA is highlighted. With proper structural parameters optimized by the PSO, the light propagates into the truncated pyramid or truncated cone microstructures in different directions, suffering little from TIR, then is refracted from sapphire into the air directly or after several reflections.

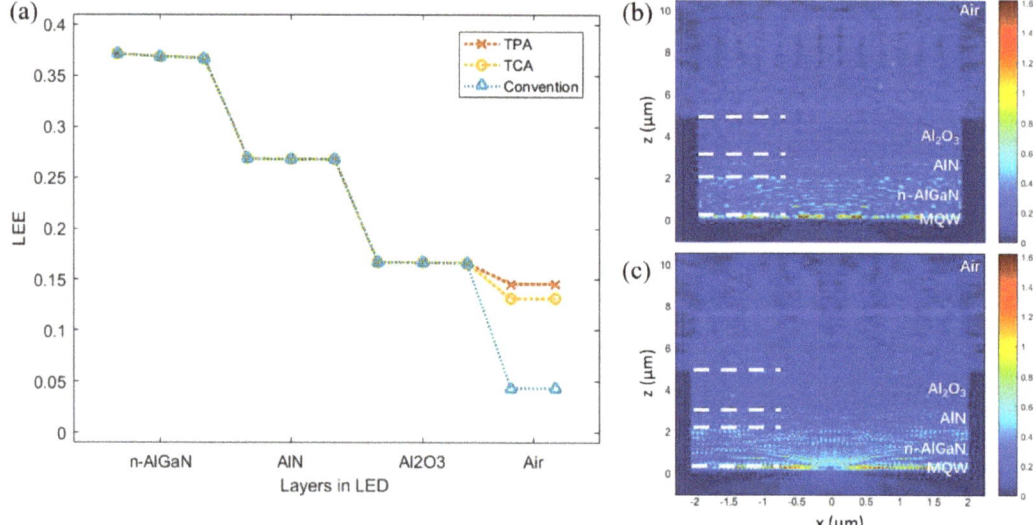

Figure 4. (a) The comparison of LEE decrement trends along LED layers, and the field distributions in the LED cross section with (b) TPA and (c) TCA.

The far-field scattering angular distribution for a specified wavelength of 280 nm is shown in Figure 5. It is calculated outside of a box-like closed surface constructed by monitors placed in homogeneous material by 3D-FDTD, according to the surface equivalence theory. Identical settings are taken to simulate the far-field distributions of the LEDs with TPA and TCA. In the conventional LED, without any light extraction structures, the main energy of the extracted light is located within the angle range (60°, 120°) owing to the TIR at the sapphire–air interface. There is still a small amount of light extracted from other angles. It is noticed that the intensity of the extracted light is largely increased with the surface structure of TPA or TCA. With the TCA, the extracted light intensity has been increased by ~1.5 folds at (75°, 115°), and ~2 folds at all other angles. With TPA, the extracted light at (75°, 115°) is less than that with TCA, but it extracts much more light at other angles, and the total power extracted by TPA is 11% more than that by TCA. There are no gaps or lacunas in the base of the TPA, which brings about collecting light at a larger angle range and larger LEE. However, above that, the four flat side walls in the truncated pyramid microstructures can cause more TIR than the truncated cone microstructure, which limits the amount of light extracted at (75°, 115°).

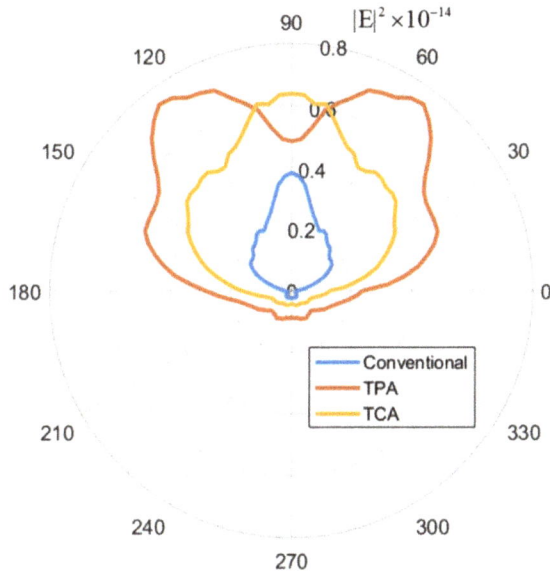

Figure 5. Far-field distribution of the LEDs with three structures.

This work is focused on the LEE enhancement in DUV-LEDs with TPA and TCA. With the help of an intelligent algorithm, we have met a clear enhancement in the LEE. Nevertheless, neither the TPA nor TCA may not be the best option for light extraction; more forms of microstructures, not only the ones located on the sapphire-air surface, can be designed and tested with the same method. The number of parameters can be expanded in order to represent more complicated microstructures.

4. Conclusions

In summary, maximizing the LEE has been the major goal for the AlGaN-based DUV-LEDs design. Here, we propose a novel scheme to cope with the main difficulty in the optimal design process as follows: consuming too much time in a parametric sweep. Drawing help from the intelligent algorithm PSO, the number of simulations in this work has been significantly reduced, and the whole process has become much more efficient. As a result, with the light extraction structures TPA and TCA, a 221% and a 257% LEE enhancement are realized at the DUV wavelength of 280 nm when compared to the conventional LEDs. The 3D-FDTD method is utilized to verify the optical properties. This method shows advantages in designing tasks that contain multiple parameters nonlinearly affecting the goal in DUV-LEDs, and it is expected to provide a new path for LED device design.

Author Contributions: Conceptualization, H.L. and T.Y.; methodology, Z.L., Y.Z. and Y.T.; software, Z.L.; validation, Z.L., Y.Z. and Y.T.; formal analysis, Z.L.; investigation, Y.Z.; resources, H.L., J.W. and T.Y.; data curation, Z.L. and Y.Z.; writing—original draft preparation, Z.L.; writing—review and editing, Y.Z., H.L. and J.W.; visualization, Z.L.; supervision, H.L. and J.W.; project administration, H.L.; funding acquisition, H.L. All authors have read and agreed to the published version of the manuscript.

Funding: This research was funded by the Guangdong Basic and Applied Basic Research Foundation, grant number 2021B1515120086; the Scientific and Technological Innovation Foundation of Foshan, grant number BK20BF013.

Data Availability Statement: The data presented in this study are available on request from the corresponding author.

Conflicts of Interest: The authors declare no conflict of interest.

References

1. Khan, A.; Balakrishnan, K.; Katona, T. Ultraviolet Light-Emitting Diodes Based on Group Three Nitrides. *Nat. Photonics* **2008**, *2*, 77–84. [CrossRef]
2. Kneissl, M.; Kolbe, T.; Chua, C.; Kueller, V.; Lobo, N.; Stellmach, J.; Knauer, A.; Rodriguez, H.; Einfeldt, S.; Yang, Z.; et al. Advances in Group III-Nitride-Based Deep UV Light-Emitting Diode Technology. *Semicond. Sci. Technol.* **2011**, *26*, 014036. [CrossRef]
3. Vilhunen, S.; Särkkä, H.; Sillanpää, M. Ultraviolet Light-Emitting Diodes in Water Disinfection. *Environ. Sci. Pollut. Res.* **2009**, *16*, 439–442. [CrossRef] [PubMed]
4. Li, J.; Gao, N.; Cai, D.; Lin, W.; Huang, K.; Li, S.; Kang, J. Multiple Fields Manipulation on Nitride Material Structures in Ultraviolet Light-Emitting Diodes. *Light Sci. Appl.* **2021**, *10*, 129. [CrossRef]
5. Guo, Y.; Yan, J.; Zhang, Y.; Wang, J.; Li, J. Enhancing the Light Extraction of AlGaN-Based Ultraviolet Light-Emitting Diodes in the Nanoscale. *J. Nanophoton.* **2018**, *12*, 043510. [CrossRef]
6. Du, P.; Zhang, Y.; Rao, L.; Liu, Y.; Cheng, Z. Enhancing the Light Extraction Efficiency of AlGaN LED with Nanowire Photonic Crystal and Graphene Transparent Electrode. *Superlattices Microstruct.* **2019**, *133*, 106216. [CrossRef]
7. Du, P.; Cheng, Z. Enhancing Light Extraction Efficiency of Vertical Emission of AlGaN Nanowire Light Emitting Diodes With Photonic Crystal. *IEEE Photonics J.* **2019**, *11*, 1600109. [CrossRef]
8. Wang, H.; Fu, L.; Lu, H.M.; Kang, X.N.; Wu, J.J.; Xu, F.J.; Yu, T.J. Anisotropic Dependence of Light Extraction Behavior on Propagation Path in AlGaN-Based Deep-Ultraviolet Light-Emitting Diodes. *Opt. Express* **2019**, *27*, A436. [CrossRef]
9. Wang, S.; Dai, J.; Hu, J.; Zhang, S.; Xu, L.; Long, H.; Chen, J.; Wan, Q.; Kuo, H.; Chen, C. Ultrahigh Degree of Optical Polarization above 80% in AlGaN-Based Deep-Ultraviolet LED with Moth-Eye Microstructure. *ACS Photonics* **2018**, *5*, 3534–3540. [CrossRef]
10. Ryu, H.-Y.; Choi, I.-G.; Choi, H.-S.; Shim, J.-I. Investigation of Light Extraction Efficiency in AlGaN Deep-Ultraviolet Light-Emitting Diodes. *Appl. Phys. Express* **2013**, *6*, 062101. [CrossRef]
11. Zhang, G.; Shao, H.; Zhang, M.; Zhao, Z.; Chu, C.; Tian, K.; Fan, C.; Zhang, Y.; Zhang, Z.-H. Enhancing the Light Extraction Efficiency for AlGaN-Based DUV LEDs with a Laterally over-Etched p-GaN Layer at the Top of Truncated Cones. *Opt. Express* **2021**, *29*, 30532. [CrossRef] [PubMed]
12. Sun, W.C.; Hsu, B.; Wei, M.K. Micro-Truncated Cone Arrays for Light Extraction of Organic Light-Emitting Diodes. In *TMS 2016 145th Annual Meeting & Exhibition*; Springer: Cham, Switzerland, 2016; pp. 473–479. [CrossRef]
13. Inoue, S.; Tamari, N.; Taniguchi, M. 150 MW Deep-Ultraviolet Light-Emitting Diodes with Large-Area AlN Nanophotonic Light-Extraction Structure Emitting at 265 Nm. *Appl. Phys. Lett.* **2017**, *110*, 141106. [CrossRef]
14. Wang, N.; Yan, W.; Qu, Y.; Ma, S.; Li, S.Z.; Qiu, M. Intelligent Designs in Nanophotonics: From Optimization towards Inverse Creation. *PhotoniX* **2021**, *2*, 22. [CrossRef]
15. Zhang, Y.; Yang, S.; Lim, A.E.-J.; Lo, G.-Q.; Galland, C.; Baehr-Jones, T.; Hochberg, M. A Compact and Low Loss Y-Junction for Submicron Silicon Waveguide. *Opt. Express* **2013**, *21*, 1310. [CrossRef]
16. Forestiere, C.; Donelli, M.; Walsh, G.F.; Zeni, E.; Miano, G.; Dal Negro, L. Particle-Swarm Optimization of Broadband Nanoplasmonic Arrays. *Opt. Lett.* **2010**, *35*, 133. [CrossRef] [PubMed]
17. Zhang, B.; Chen, W.; Wang, P.; Dai, S.; Li, H.; Lu, H.; Ding, J.; Li, J.; Li, Y.; Fu, Q.; et al. Particle Swarm Optimized Polarization Beam Splitter Using Metasurface-Assisted Silicon Nitride Y-Junction for Mid-Infrared Wavelengths. *Opt. Commun.* **2019**, *451*, 186–191. [CrossRef]
18. Shokooh-Saremi, M.; Magnusson, R. Particle Swarm Optimization and Its Application to the Design of Diffraction Grating Filters. *Opt. Lett.* **2007**, *32*, 894. [CrossRef]

19. Ryu, H.Y. Numerical Study on the Wavelength-Dependence of Light Extraction Efficiency in AlGaN-Based Ultraviolet Light-Emitting Diodes. *Opt. Quant. Electron.* **2014**, *46*, 1329–1335. [CrossRef]
20. Ryu, H.Y.; Shim, J.-I. Structural Parameter Dependence of Light Extraction Efficiency in Photonic Crystal InGaN Vertical Light-Emitting Diode Structures. *IEEE J. Quantum Electron.* **2010**, *46*, 714–720. [CrossRef]
21. Zhao, P.; Zhao, H. Analysis of Light Extraction Efficiency Enhancement for Thin-Film-Flip-Chip InGaN Quantum Wells Light-Emitting Diodes with GaN Micro-Domes. *Opt. Express* **2012**, *20*, A765. [CrossRef]

Article

Investigation into the MOCVD Growth and Optical Properties of InGaN/GaN Quantum Wells by Modulating NH₃ Flux

Zhenyu Chen [1,2], Feng Liang [1,*], Degang Zhao [1,3,*], Jing Yang [1], Ping Chen [1] and Desheng Jiang [1]

[1] State Key Laboratory of Integrated Optoelectronics, Institute of Semiconductors, Chinese Academy of Sciences, Beijing 100083, China
[2] College of Materials Science and Opto-Electronic Technology, University of Chinese Academy of Sciences, Beijing 100049, China
[3] Center of Materials Science and Optoelectronics Engineering, University of Chinese Academy of Sciences, Beijing 100049, China
* Correspondence: liangfeng13@semi.ac.cn (F.L.); dgzhao@semi.ac.cn (D.Z.)

Abstract: In this study, the surface morphology and luminescence characteristics of InGaN/GaN multiple quantum wells were studied by applying different flow rates of ammonia during MOCVD growth, and the best growth conditions of InGaN layers for green laser diodes were explored. Different emission peak characteristics were observed in temperature-dependent photoluminescence (TDPL) examination, which showed significant structural changes in InGaN layers and in the appearance of composite structures of InGaN/GaN quantum wells and quantum-dot-like centers. It was shown that these changes are caused by several effects induced by ammonia, including both the promotion of indium corporation and corrosion from hydrogen caused by the decomposition of ammonia, as well as the decrease in the surface energy of InGaN dot-like centers. We carried out detailed research to determine ammonia's mechanism of action during InGaN layer growth.

Keywords: InGaN quantum well; surface morphology; NH₃ flux; GaN lasers; MOCVD

Citation: Chen, Z.; Liang, F.; Zhao, D.; Yang, J.; Chen, P.; Jiang, D. Investigation into the MOCVD Growth and Optical Properties of InGaN/GaN Quantum Wells by Modulating NH₃ Flux. *Crystals* **2023**, *13*, 127. https://doi.org/10.3390/cryst13010127

Academic Editors: Francisco M. Morales and Dmitri Donetski

Received: 28 November 2022
Revised: 26 December 2022
Accepted: 4 January 2023
Published: 10 January 2023

Copyright: © 2023 by the authors. Licensee MDPI, Basel, Switzerland. This article is an open access article distributed under the terms and conditions of the Creative Commons Attribution (CC BY) license (https://creativecommons.org/licenses/by/4.0/).

1. Introduction

InGaN/GaN multiple quantum well-based light emitting devices have attracted a large amount of attention during the past few years, especially in the wavelength range of blue and green [1–4]. It is necessary to manufacture high-performance green InGaN/GaN active regions through metal–organic chemical vapor deposition (MOCVD), as they have promising applications in laser display and solid-state lighting [4,5]. However, when an emitting wavelength of an active region is in the green range, the quality of the active region seriously decreases due to a large increase in indium content in the InGaN layer [6].

The incorporation of indium into InGaN is sensitive to several growth conditions, such as temperature, growth rate, flux of sources, pseudo-template characteristic effect, etc. [7,8]. Ammonia is an important gas source in the MOCVD-based growth of GaN-based materials; usually, an increase in the ammonia flow rate is considered to have a positive effect on indium incorporation, because a high V/III ratio is often beneficial to the incorporation of group III atoms [9,10]. However, in this study, we found that ammonia may have a more complicated effect on growth in the InGaN/GaN active region, including on both the optical and structural properties of quantum wells. When the ammonia flow rate was increased to a certain degree, a corrosive effect of ammonia on the InGaN layer was observed. We carried out further studies to explore the interaction mechanism between ammonia and the InGaN layer and its significant influences on the properties of quantum wells. In this work, different ammonia flux during quantum well growth resulted in different emission mechanisms. Analysis on those different emission mechanisms may be beneficial for high quality green InGaN quantum well growth.

2. Experiments

Four InGaN/GaN multi-quantum-well (MQW) samples (series I, A, B, C, and D) and five InGaN/GaN single-quantum-well (SQW) samples (series II, S1, S2, and S3; and series III, R1 and R2) were grown on 2 inch c-plane (0001) sapphire substrates via a Thomas Swan MOCVD with close-coupled showerhead reactor. During MOCVD growth process, trimethylindium (TMIn) and trimethylgallium (TMGa) were employed as the metal precursors for gallium and indium, and ammonia (NH_3) acted as the nitrogen precursor. Hydrogen was used as the carrier gas for the quantum well structure (including quantum wells and quantum barriers) growth and nitrogen was used as carrier gas for all the other layers. Bicyclopentadienyl magnesium (Cp_2Mg) and silane were used for p- and n-type doping, respectively.

The epitaxial structure of the MQW samples in series I consisted of a 23 nm GaN nucleation layer; a 1.5 µm, unintentionally doped GaN layer; a 1 µm, Si-doped n-GaN layer (electron concentration = $3.5 \times 10^{18}/cm^3$); a 2-period InGaN (3 nm)/GaN (8 nm) MQW active region; a 1 µm, Mg-doped p-GaN layer (hole concentration = $2.1 \times 10^{17}/cm^3$); and a 40 nm, p++ type GaN layer. The epitaxial structure was then annealed in nitrogen gas at 800 °C for 180 s. The epitaxial structure of the SQW samples in series II did not have P-type layers, and no annealing process was carried out after the epitaxial growth was realized. The structure of these samples consisted of a 23 nm GaN nucleation layer; a 1.5 µm, unintentionally doped GaN; a 1 µm, Si-doped n-GaN layer; and an InGaN (3 nm)/GaN (8 nm) SQW. Samples in series III were grown under the same conditions as those in series II, except a p-type layer was grown above the SQW region, and afterwards, they underwent an annealing process. All the film thicknesses were controlled by post-growth analysis via HRXRD to ensure different samples in the same series have the same thickness for each layer.

The growth conditions for all 4 MQW samples in series I were the same, except for the NH_3 flow rate during quantum well growth, which was different, being 1, 2, 3, and 4 slm for samples A, B, C, and D, respectively. V/III-ratios of these samples were 6800, 13,600, 20,400 and 27,200 corresponding to the different ammonia flow rates. The SQW samples in series II were grown under the same conditions as those in series I, and the NH_3 flow rate was 1, 2, and 4 slm for samples S1, S2, and S3, respectively. The samples in series III were grown under different NH_3 flow rates: 2 and 4 slm for samples R1 and R2. The schematic diagrams for the InGaN/GaN MQW and SQW samples are shown in Figure 1, and the growth conditions are shown in Table 1. In this work, modulating ammonia flux did not result in an obvious growth rate change. Samples in the series I share similar film thickness for each layer as results of HRXRD shown.

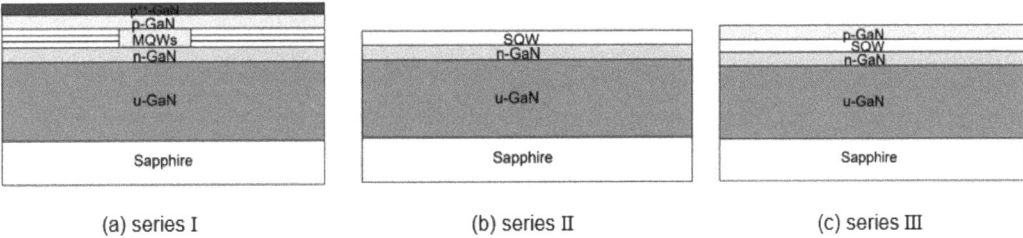

Figure 1. Schematic diagrams for InGaN/GaN MQW (series I) and SQW (series II/III) samples.

The samples in series I were characterized via high resolution X-ray diffraction (HRXRD), electroluminescence (EL) spectroscopy and temperature-dependent photoluminescence (TDPL) spectroscopy measurements. An in-plane scan was performed via HRXRD, employing a Cu Kα1 line. The EL spectra were recorded using a high-resolution spectrometer using a direct current injection. The TDPL spectra were recorded from 30K to 300K using a 405 nm GaN semiconductor laser as the excitation source. Micro-PL images were taken under a confocal microscope with a 405 nm GaN semiconductor laser as the excitation source. Samples in series II and III were analyzed using an atomic force microscope (AFM) in a surface morphology test.

Table 1. Quantum well growth conditions for different samples of series I, II, and III with varying ammonia flow rates.

Series	Sample Name	Temperature (°C)	NH$_3$ Flux (slm)
I	A	665	1
	B	665	2
	C	665	3
	D	665	4
II	S1	665	1
	S2	665	2
	S3	665	4
III	R1	665	2
	R2	665	4

3. Results and Discussions

The HRXRD results for the samples in series I showed that in the four samples, the quantum wells and quantum barriers had the same thicknesses (3 nm and 8 nm, respectively). The electroluminescence (EL) spectra of the samples in series I were measured, and the results are presented in Figure 2. As the current injection increased, the peak wavelengths of all of the samples showed a decreasing trend, as shown in Figure 2a. First, the wavelength of the EL peak increased when the NH$_3$ flow rate was below 2 slm during the InGaN layer growth, but the peak wavelength decreased with a further increase when the flow rate was over 2 slm, which indicated that the indium content in InGaN quantum wells decreases when the NH$_3$ flow rate is increased over a certain threshold, as shown in Figure 2b. Previous research has shown that NH$_3$ may have both positive and negative effects on the incorporation of indium atoms into InGaN quantum wells in the wavelength range of blue LED [11,12]. It has been reported that when InGaN is generated in MOCVD by NH$_3$, TMIn, and TMGa, an additional reaction of $NH_3 - NH_3 \rightarrow (1-x)NH_3 + x\left(\frac{1}{2}N_2 + \frac{3}{2}H_2\right)$—occurs [13]. The NH$_3$ dissociation reaction provides more H$_2$, and it is proved to have a corrosive effect on indium atoms [14,15]. For green InGaN/GaN MQW LDs, the incorporation of indium can be more sensitive to growth conditions due to the higher indium content in InGaN. For the samples in series I, the EL result corresponded well with results that have been previously reported [11,12].

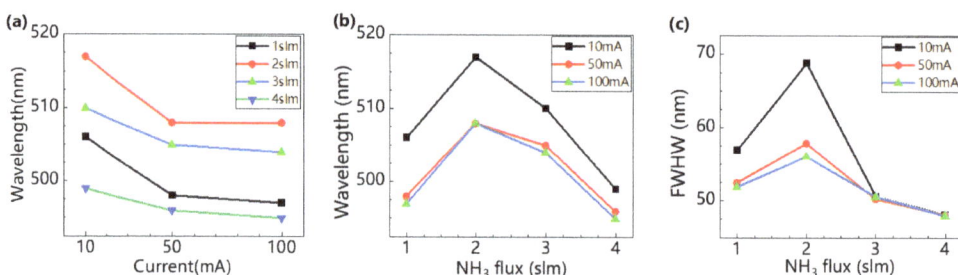

Figure 2. EL peak wavelength vs. injection current and NH$_3$ flux (a,b); FWHM (c) dependence on varying NH$_3$ flux measured under different injections of 10, 50 and 100 mA. The lines are connected for visual purposes.

However, a strange increase in the FWHM of the EL spectral peak (10~20 nm higher than other samples) was observed in sample B (grown at 2 slm NH$_3$ flux), as shown in

Figure 2c. Additionally, a remarkable decrease in the peak wavelength of the EL spectra occurred when the injection current increased from 10 to 50 mA, as shown in Figure 2b. The decreases in samples A and B (approximately 8–9 nm) were a little larger than those in samples C and D. These phenomena indicate that ammonia flux could influence the incorporation of indium into InGaN layers. However, more effects may be caused by the change in NH_3 flux during growth. A set of PL/TDPL measurements of samples A, B, and D were made to characterize the emitting mechanism of InGaN/GaN MQW samples with different NH_3 flow rates, and this is discussed in detail below.

The TDPL spectra data are shown in Figure 3. We attributed the small, undulating peaks observed in Figure 3a–c to interference fringes due to the Fabry–Perot effect of the epitaxial films, and Gaussian fitting was applied to obtain the spectral line shape of the luminescence peak and eliminate the disturbance of interference. The wavelengths of emission peaks shown in Figure 3d–f were collected from Gaussian fitting peaks in Figure 3a–c, respectively. Compared to the EL results in Figure 2, we observed that the TDPL spectra showed totally different luminescence modes in these samples. Because EL testing has a much higher carrier injection than TDPL, the emission peaks in EL spectra were mainly derived from quantum wells, as the intensity of the emission peaks from localized states was much weaker compared to that from quantum wells under high injection conditions. In comparison, TDPL spectra show more information regarding those peaks which may be neglected under high injection conditions. Different luminescence modes were found in samples A, B, and D. In sample B, two main emission peaks were observed, but in samples A and D, only one emission peak was observed. A reasonable explanation for this is that there might have been two different luminescence mechanisms in these three samples, i.e., emissions from QD-like (quantum-dot-like) centers and from quantum wells. A typical emission peak from InGaN QD-like centers has characteristics which include a low spectral width of emission peaks, high intensity at a low injection rate, and low sensitivity to temperature change [16]. For emissions from quantum wells, a larger width of emission peaks and a more sensitive temperature dependence is expected. In sample B, two different emission peaks, PM and PD, were observed. The green emission peak (at 515 nm at 30K, i.e., PM in Figure 3b) had a broader width over 60 nm, and the intensity of this peak decreased sharply with the increasing temperature, especially at the room temperature range, in which this peak was too weak to be detected. Because of the weak emission when the temperature was over 150K, the FWHM and wavelength of this peak may lack accuracy, and the analysis on this peak were all based on the data collected before 150K. Another yellow peak (at 590 nm at 30K, i.e., PD in Figure 3b) had greater intensity, a smaller width around 40 nm, and was more stable when the temperature increased. Thus, these two peaks could be attributed to two different emission origins: the green peak to quantum well emission and the yellow peak to QD-like center emission. When sample A was examined, a high-intensity peak with a small width under 40 nm was detected. Additionally, there was only a slight decrease in the peak intensity with the increase in temperature. The change tendency of the peak wavelength and FWHM (Figure 4a,b) was similar to that of the PD peak in sample B, as the peak wavelength shifting was under 3 nm and the FWHM was lower than 50 nm, indicating that these two peaks should have had similar luminescence mechanisms. Correspondingly, an emission peak with a larger FWHM around 65 nm was found in sample D, and the intensity of this peak decreased rapidly when temperature increased, especially when it increased to the room temperature range (not shown in Figure 3f). The origin of this peak should have been similar to PM. The peak wavelengths of PM and the peak in sample D both had a 5 nm increase when the temperature rose from 30 K to 100 K, and both had a decrease trend when the temperature was over 100 K. However, the temperature dependence of the peak wavelength of PM was quite different from the peak in sample D when the temperature was over 150 K, as can be seen in Figure 3e,f. We think the reason for this could be that the PL peak intensity of PM and sample D decreased rapidly with the increase in temperature, especially when the temperature was over 150 K. As a result, the peak wavelength of these two peaks could be severely affected by the adjacent peaks, which became clearly stronger. Thus, we determined that in sample A, only QD-like center emissions existed, and in sample D, only quantum well emissions existed. Additionally, the characteristics of the TDPL emission peaks from sample D and PM in sample B under room temperature conditions were consistent with the EL results shown in Figure 2. It can be concluded that in these two cases, the emission mechanism was the same as that in EL testing, in which emissions were derived from the InGaN quantum well matrix. However, there were no such emission peaks in sample A under TDPL, which was consistent with EL spectra. A reasonable explanation for this is that more deep localized states existed in sample A, and then, carrier recombination in QD-like structures predominated under a small injection rate. In comparison, carrier recombination mainly took place in the quantum well matrix under a high injection rate in EL. As a result, the intensity of emissions from QD-like structures was much higher than that from the quantum well matrix under small injection rates, which made it difficult to observe the emissions from the quantum well matrix in TDPL. Additionally, in terms of EL testing, only emissions from the quantum well matrix could be observed, which was the same as that in sample B and D.

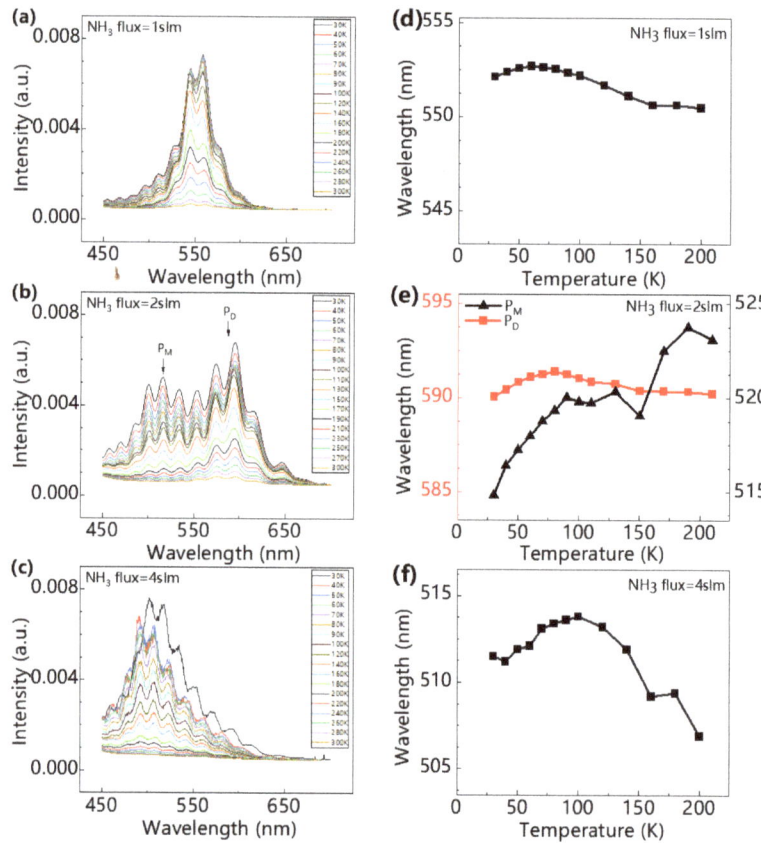

Figure 3. TDPL spectra (**a**–**c**) and temperature-dependent emission peak wavelength in series II (**d**–**f**) for samples A, B, and D, grown with NH$_3$ flux of 1, 2, and 4 slm, respectively. In Figure 3 (**b**,**e**), PD is assigned to the peak from quantum-dot-like centers, and PM is assigned to the peak from quantum wells. The connected lines in (**d**–**f**) are shown for visual purposes.

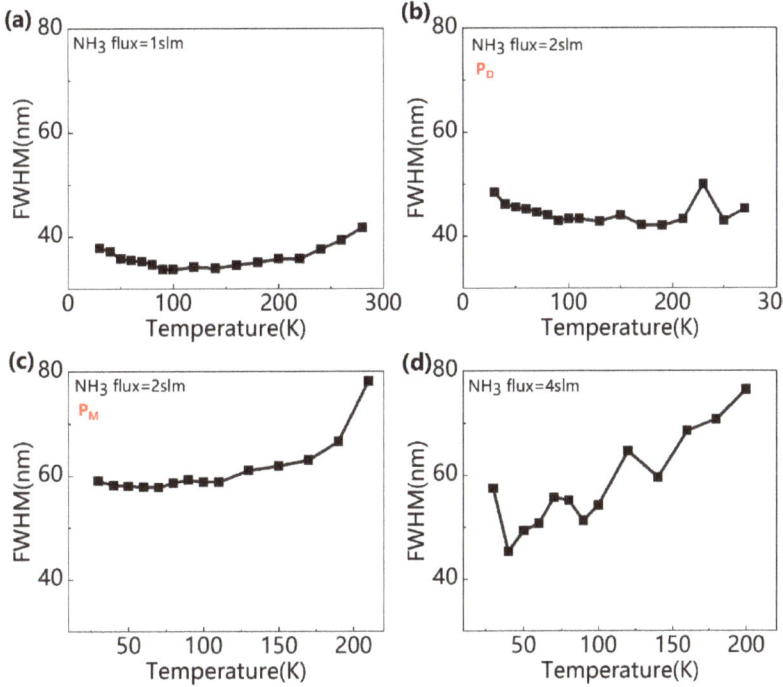

Figure 4. Temperature-dependent full width at half magnitude (FWHM) of PL peaks for samples A, B, and D in series II (**a**–**d**) grown with NH$_3$ flux of 1, 2, and 4 slm, respectively, where PD is marked for the peak from QD-like centers and PM is marked for the peak from quantum wells (**b**,**c**). The connected lines are only used for visual purposes.

In order to further examine whether there was a structure in these QWs which may have supplied a composite luminescence mechanism for both QW and QD-like emissions, the samples in series II and series III were tested.

In series II, samples S1, S2, and S3 were prepared based on the growth conditions of samples A, C, and D, respectively. Instead of two-period-MQWs, only a single quantum well was grown in these samples. AFM images of samples in series II are shown in Figure 5, and the height distribution at a cross section of the surface near white spots in Figure 5 is correspondingly shown in Figure 6.

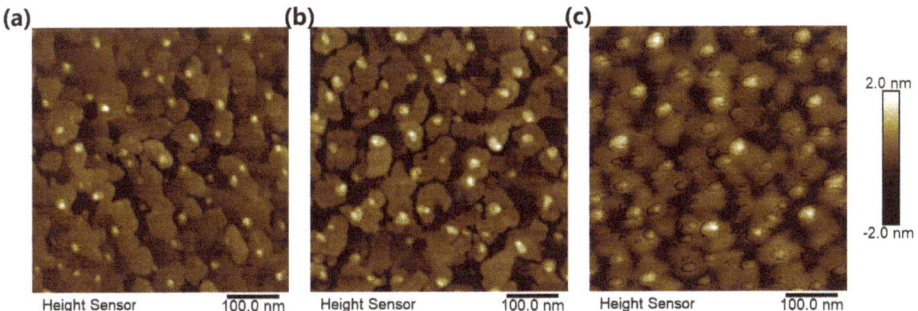

Figure 5. AFM images for samples S1 (**a**), S2 (**b**), and S3 (**c**), where white spots are assigned as QD–like centers.

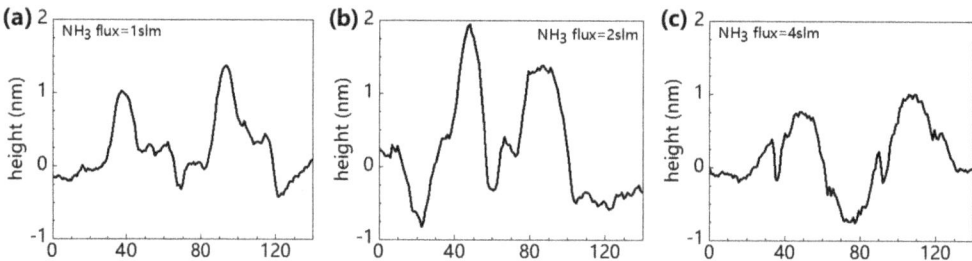

Figure 6. Height distribution at a cross section of the surface near white spots in AFM results for samples S1 (**a**), S2 (**b**), and S3 (**c**), correspondingly.

In all of these three samples, the surface topography suggested Stranski–Krastanov mode growth had a remarkable influence, which allows a 3D nanostructure to grow on a 2D wetting layer [17]. For sample S1 and sample S2, the InGaN epilayers were mainly composed of three parts, i.e., a 2D wetting layer, a 3D large island layer, and large amounts of QD-like centers located above the large islands. However, in sample S3, only mild fluctuations or thick dots could be found on the surface, which looked like flat island aggregates. AFM data show that in comparison to sample S1, sample S2 had QD-like centers with higher densities and larger sizes. The QD-like structures in sample S3 were much shorter and thicker than those in sample S1 and S2 as Table 2 shown.

Table 2. Characteristics of QD-like centers in samples S1, S2, and S3.

Sample	Density/cm^{-2}	Diameter/nm	Height/nm
S1	1.9×10^{10}	15~20	1~2
S2	2.7×10^{10}	20~30	1~2
S3	2.7×10^{10}	35~40	~1

According to a previous report [18], green InGaN/GaN MQW samples often show morphologies with large amounts of islands on the GaN layer with QDs above these islands. The formation of InGaN QDs during epitaxial growth is based on the typical Stranski–Krastanov growth mode [18]. Due to lattice mismatch between InGaN and GaN, the InGaN layer grown on GaN will suffer from a high level of compressive stress, and the strain energy will accumulate during the growth of the InGaN layer. When the thickness of the InGaN layer becomes more than a critical thickness, the transition from 2D to 3D growth mode will occur. Additionally, when the thickness of the InGaN layer becomes even higher, it will meet another critical thickness level at which layer relaxation occurs. Thus, strained QD-like centers will form on the top of the InGaN layer. In addition, the critical thickness of the wetting layer for the formation of QD-like centers decreases when the indium composition of the InGaN layer increases [19,20]. Kobayashi reported an S-K growth mode InGaN layer of 50% indium without the formation of QDs on SiC substrate [21], but it was also reported that strained QDs were observed after a 0.9 monolayer (ML) was grown when growing InN on a GaN substrate [22,23]. We think it may be possible for the growth mode to change from the S-K growth mode to a complete V-W growth mode when the indium composition in InGaN increases from 50% to 100%. It should also be mentioned that when using MBE to epitaxially grow the InGaN layer, the critical thickness of the wetting layer for QD's formation is smaller than that of MOCVD [24,25]. In fact, different surface structures and dynamic processes of MBE and MOCVD will lead to different surface energies, which may result in differences in the critical thicknesses.

Referring to the observed results for samples in series II, ammonia flux during growth had a significant effect on the size and density of QD-like centers. The structural characteristics of QD-like centers in series II samples are listed in Table 2. The density data were calculated by counting white spots in a unit area, and the height and diameter data were obtained by averaging five height and diameter figures measured from cross section images for each sample. The mechanism of how ammonia flux affects the formation of QD-like centers can be explained from two perspectives.

Firstly, in terms of surface thermodynamics, it was found that when growing InAs self-assembled QDs on the InP substrate, the size of InAs QDs will decrease as the AsH_3 flow rate decreases [26]. A lower flow rate will result in a change in the pressure condition, which makes the surface energy of InAs increase. Higher surface energy supports the formation of smaller-size QDs. Additionally, such a chemical process is also effective in the formation of InGaN QD-like centers in MOCVD.

Secondly, in terms of surface dynamics, when growing GaN using MBE, a higher flow rate of nitride will increase the surface diffusion barrier of gallium atoms [27]. Oliver et al. [28] reported that the diffusion of adatoms on the surface was affected not only by active nitrogen atoms, but also by hydrogen radicals. Hydrogen radicals can be found in the decomposition of ammonia during MOCVD epitaxy. Increasing the flow rate of ammonia or promoting the decomposition of ammonia can decrease the surface diffusion barrier of adatoms, therefore increasing the mobility of adatoms. In conclusion, increasing the ammonia flow rate can increase the size of InGaN QD-like centers by decreasing their surface energy and enhancing the diffusion mobility of indium adatoms.

As a result of these two factors, it was shown that sample S1 had a morphology with smaller QD-like structures, which was more efficient in confining carriers in deep localized states, generating strong emissions with long wavelengths, as presented in TDPL spectra. When the QD-like centers grew thicker with higher ammonia flux, especially over 2 slm, in sample S2 and S3, carriers could not be confined well in such QD-like centers. Thus, more carrier combination took place in the quantum well matrix, leading to weaker emissions at long wavelengths and stronger emissions at short wavelengths, as shown in the TDPL results. Additionally, the characteristics of those stronger emissions at short wavelengths corresponded well with the emission peaks detected in EL testing, which mainly originated in radiative recombination in the quantum well matrix.

In the series III samples, where QD-like centers were formed under a GaN cap layer and underwent an annealing procedure, QD-like centers tended to combine with each other, which made the size of QD-like centers larger, together with a decrease in their density, which can be seen in Figure 7. In sample R1 (Figure 7b), the islands became enhanced as QD-like centers merged with each other. In sample R2 (Figure 7d), the homogeneity of QD-like centers and the layer under them seemed to clearly decrease. Additionally, we speculated that too-large QD-like centers may restrain the QD-like center emissions after the annealing process.

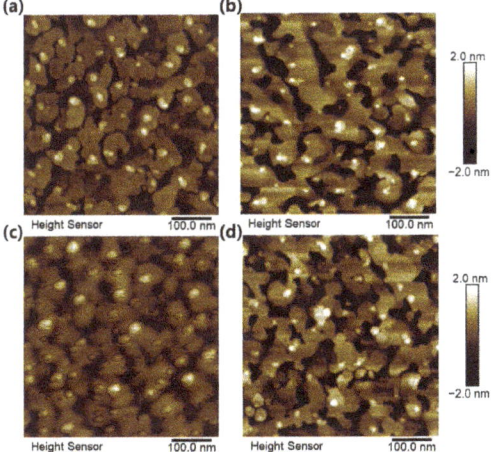

Figure 7. AFM images of sample S1 (**a**), R1 (**b**), S2 (**c**), and R2 (**d**), where white spots are QD−like centers.

4. Conclusions

In this study, the effects of ammonia flux on the MOCVD epitaxial growth of the InGaN layer and the properties of related InGaN/GaN MQW (or SQW) structures were investigated. The results can be summarized to two aspects. Firstly, ammonia has great influence on the incorporation of indium atoms into InGaN, especially the structural and emission properties of quantum wells. A higher ammonia flow rate leads to a larger V/III ratio, resulting in high nitrogen atom concentration, which is beneficial to the growth of InGaN. With higher nitrogen atom concentration, indium atoms are more likely to incorporate into the InGaN epilayer, resulting in higher indium content in InGaN quantum wells. However, when the ammonia flow rate increases beyond a threshold in which the disassociation of ammonia plays a dominant role, the corrosion of indium caused by extra H_2 generated from ammonia decomposition will decrease the indium content in InGaN, resulting in a blue shift of the emission peak.

It is noted that ammonia has an obvious effect on the surface morphology of the InGaN epilayer. With an ammonia flow rate over 2 slm, surface atoms have higher migration mobility, and the formation of QD-like centers has lower surface energy. These factors lead to thicker and shorter QD-like centers, which are less effective in carrier localization. Based on different conditions of the InGaN layer surface, InGaN MQWs will show two different luminescence mechanisms. In TDPL testing, samples with low ammonia flux presented typical emission peaks of QD-like centers, and samples with high ammonia flux showed typical emission peaks of quantum wells. However, all of the samples showed the emission peaks of quantum wells in EL testing.

Emissions from QD-like centers are remarkable under small injection rates, such as that in PL testing, but when it comes to large injection rates, as in laser diodes, the emissions from the quantum well matrix dominate. Further research is needed to determine whether it is feasible to utilize the emissions from QD-like centers for green laser diodes. However, if we want to optimize the emissions from the quantum well matrix, it is possible to eliminate the effect of QD-like centers by increasing ammonia flux during MOCVD growth. The results of this study regarding the effect of ammonia flux during InGaN MOCVD growth show that it is more promising to grow epitaxially high-performance green laser diodes with InGaN/GaN MQWs.

Author Contributions: Formal analysis, Z.C., F.L. and D.Z.; investigation, Z.C. and F.L.; resources, F.L. and D.Z.; writing—original draft preparation, Z.C. and F.L.; writing—review and editing, D.J., J.Y. and P.C.; supervision, D.Z.; project administration, D.Z.; funding acquisition, F.L., D.Z. and J.Y. All authors have read and agreed to the published version of the manuscript.

Funding: This work was supported by National Key R&D Program of China (Grant No. 2022YFB3608100), Beijing Municipal Science & Technology Commission, Administrative Commission of Zhongguancun Science Park (Z211100007921022, Z211100004821001), National Natural Science Foundation of China (Grant Nos. 62034008, 62074142, 62074140, 61974162, 61904172, 61874175, 62127807, U21B2061), Key Research and Development Program of Jiangsu Province (BE2021008-1), Beijing Nova Program (Grant No. 202093), Strategic Priority Research Program of Chinese Academy of Sciences (Grant No. XDB43030101), and Youth Innovation Promotion Association of Chinese Academy of Sciences (Grant No. 2019115).

Institutional Review Board Statement: Not applicable.

Informed Consent Statement: Not applicable.

Data Availability Statement: The data that support the findings of this study are available from the corresponding author upon reasonable request.

Conflicts of Interest: The authors declare no conflict of interest.

References

1. Baten, M.Z.; Alam, S.; Sikder, B.; Aziz, A. III-Nitride Light-Emitting Devices. *Photonics* **2021**, *8*, 430. [CrossRef]
2. Sizov, D.; Bhat, R.; Zah, C.E. Gallium Indium Nitride-Based Green Lasers. *J. Light. Technol.* **2012**, *30*, 679–699. [CrossRef]

3. Liang, F.; Zhao, D.; Liu, Z.; Chen, P.; Yang, J.; Duan, L.; Shi, Y.; Wang, H. GaN-based blue laser diode with 6.0 W of output power under continuous-wave operation at room temperature. *J. Semicond.* **2021**, *42*, 112801. [CrossRef]
4. Yang, J.; Zhao, D.; Liu, Z.; Liang, F.; Chen, P.; Duan, L.; Wang, H.; Shi, Y. A 357.9 nm GaN/AlGaN multiple quantum well ultraviolet laser diode. *J. Semicond.* **2022**, *43*, 010501. [CrossRef]
5. Zhi, T.; Tao, T.; Liu, X.; Xue, J.; Wang, J.; Tao, Z.; Li, Y.; Xie, Z.; Liu, B. Low-threshold lasing in a plasmonic laser using nanoplate InGaN/GaN. *J. Semicond.* **2021**, *42*, 122803. [CrossRef]
6. Liang, Y.; Liu, J.; Ikeda, M.; Tian, A.; Zhou, R.; Zhang, S.; Liu, T.; Li, D.; Zhang, L.; Yang, H. Effect of inhomogeneous broadening on threshold current of GaN-based green laser diodes. *J. Semicond.* **2019**, *40*, 052802. [CrossRef]
7. Wang, X.W.; Liang, F.; Zhao, D.G.; Liu, Z.S.; Zhu, J.J.; Peng, L.Y.; Yang, J. Improving the homogeneity and quality of InGaN/GaN quantum well exhibiting high in content under low TMIn flow and high pressure growth. *Appl. Surf. Sci.* **2021**, *548*, 10. [CrossRef]
8. Benzarti, Z.; Sekrafi, T.; Bougrioua, Z.; Khalfallah, A.; El Jani, B. Effect of SiN Treatment on Optical Properties of In(x)Ga1-x N/GaN MQW Blue LEDs. *J. Electron. Mater.* **2017**, *46*, 4312–4320. [CrossRef]
9. Kim, S. The influence of ammonia pre-heating to InGaN films grown by TPIS-MOCVD. *J. Cryst. Growth* **2003**, *247*, 55–61. [CrossRef]
10. Kunzmann, D.J.; Kohlstedt, R.; Uhlig, T.; Schwarz, U.T. Critical discussion of the determination of internal losses in state-of-the-art (Al,In)GaN laser diodes. In Proceedings of the Conference on Gallium Nitride Materials and Devices XV, San Francisco, CA, USA, 4–6 February 2020.
11. Yang, J.; Zhao, D.G.; Jiang, D.S.; Chen, P.; Zhu, J.J.; Liu, Z.S.; Liu, W.; Liang, F.; Li, X.; Liu, S.T.; et al. Increasing the indium incorporation efficiency during InGaN layer growth by suppressing the dissociation of NH_3. *Superlattices Microstruct.* **2017**, *102*, 35–39. [CrossRef]
12. Yang, J.; Zhao, D.G.; Jiang, D.S.; Chen, P.; Zhu, J.J.; Liu, Z.S.; Liu, W.; Li, X.; Liang, F.; Liu, S.T.; et al. Investigation on the corrosive effect of NH_3 during InGaN/GaN multi-quantum well growth in light emitting diodes. *Sci. Rep.* **2017**, *7*, 44850. [CrossRef] [PubMed]
13. Hirasaki, T.; Hasegawa, T.; Meguro, M.; Thieu, Q.T.; Murakami, H.; Kumagai, Y.; Monemar, B.; Koukitu, A. Investigation of NH_3 input partial pressure for N-polarity InGaN growth on GaN substrates by tri-halide vapor phase epitaxy. *Jpn. J. Appl. Phys.* **2016**, *55*, 05FA01. [CrossRef]
14. Czernecki, R.; Kret, S.; Kempisty, P.; Grzanka, E.; Plesiewicz, J.; Targowski, G.; Grzanka, S.; Bilska, M.; Smalc-Koziorowska, J.; Krukowski, S.; et al. Influence of hydrogen and TMIn on indium incorporation in MOVPE growth of InGaN layers. *J. Cryst. Growth* **2014**, *402*, 330–336. [CrossRef]
15. Czernecki, R.; Grzanka, E.; Smalc-Koziorowska, J.; Grzanka, S.; Schiavon, D.; Targowski, G.; Plesiewicz, J.; Prystawko, P.; Suski, T.; Perlin, P.; et al. Effect of hydrogen during growth of quantum barriers on the properties of InGaN quantum wells. *J. Cryst. Growth* **2015**, *414*, 38–41. [CrossRef]
16. Wang, L.; Wang, L.; Yu, J.; Hao, Z.; Luo, Y.; Sun, C.; Han, Y.; Xiong, B.; Wang, J.; Li, H. Abnormal Stranski-Krastanov Mode Growth of Green InGaN Quantum Dots: Morphology, Optical Properties, and Applications in Light-Emitting Devices. *ACS Appl. Mater. Interfaces* **2019**, *11*, 1228–1238. [CrossRef]
17. Kour, R.; Arya, S.; Verma, S.; Singh, A.; Mahajan, P.; Khosla, A. Review—Recent Advances and Challenges in Indium Gallium Nitride (InxGa1-xN) Materials for Solid State Lighting. *ECS J. Solid State Sci. Technol.* **2019**, *9*, 015011. [CrossRef]
18. Pristovsek, M.; Kadir, A.; Meissner, C.; Schwaner, T.; Leyer, M.; Stellmach, J.; Kneissl, M.; Ivaldi, F.; Kret, S. Growth mode transition and relaxation of thin InGaN layers on GaN (0001). *J. Cryst. Growth* **2013**, *372*, 65–72. [CrossRef]
19. Pristovsek, M.; Stellmach, J.; Leyer, M.; Kneissl, M. Growth mode of InGaN on GaN (0001) in MOVPE. *Phys. Status Solidi (C)* **2009**, *6*, S565–S569. [CrossRef]
20. Leyer, M.; Stellmach, J.; Meissner, C.; Pristovsek, M.; Kneissl, M. The critical thickness of InGaN on (0001)GaN. *J. Cryst. Growth* **2008**, *310*, 4913–4915. [CrossRef]
21. Nakatsu, Y.; Nagao, Y.; Hirao, T.; Hara, Y.; Masui, S.; Yanamoto, T.; Nagahama, S.I.; Morkoç, H.; Fujioka, H.; Schwarz, U.T. Blue and green InGaN semiconductor lasers as light sources for displays. In *Gallium Nitride Materials and Devices XV*; SPIE: Bellingham, WA, USA, 2020.
22. Meissner, C.; Ploch, S.; Pristovsek, M.; Kneissl, M. Volmer-Weber growth mode of InN quantum dots on GaN by MOVPE. *Phys. Status Solidi (C)* **2009**, *6*, S545–S548. [CrossRef]
23. Ivaldi, F.; Meissner, C.; Domagala, J.; Kret, S.; Pristovsek, M.; Högele, M.; Kneissl, M. Influence of a GaN Cap Layer on the Morphology and the Physical Properties of Embedded Self-Organized InN Quantum Dots on GaN(0001) Grown by Metal–Organic Vapour Phase Epitaxy. *Jpn. J. Appl. Phys.* **2011**, *50*, 031004. [CrossRef]
24. Grandjean, N.; Massies, J. Real time control of InxGa1−xN molecular beam epitaxy growth. *Appl. Phys. Lett.* **1998**, *72*, 1078–1080. [CrossRef]
25. Adelmann, C.; Simon, J.; Feuillet, G.; Pelekanos, N.T.; Daudin, B.; Fishman, G. Self-assembled InGaN quantum dots grown by molecular-beam epitaxy. *Appl. Phys. Lett.* **2000**, *76*, 1570–1572. [CrossRef]
26. Alghoraibi, I.; Rohel, T.; Bertru, N.; Le Corre, A.; Létoublon, A.; Caroff, P.; Dehaese, O.; Loualiche, S. Self-assembled InAs quantum dots grown on InP (3 1 1)B substrates: Role of buffer layer and amount of InAs deposited. *J. Cryst. Growth* **2006**, *293*, 263–268. [CrossRef]

27. Jiang, L.; Liu, J.; Zhang, L.; Qiu, B.; Tian, A.; Hu, L.; Li, D.; Huang, S.; Zhou, W.; Ikeda, M.; et al. Suppression of substrate mode in GaN-based green laser diodes. *Opt. Express* **2020**, *28*, 15497–15504. [CrossRef] [PubMed]
28. Oliver, R.A.; Kappers, M.J.; Humphreys, C.J.; Briggs, G.A.D. The influence of ammonia on the growth mode in InGaN/GaN heteroepitaxy. *J. Cryst. Growth* **2004**, *272*, 393–399. [CrossRef]

Disclaimer/Publisher's Note: The statements, opinions and data contained in all publications are solely those of the individual author(s) and contributor(s) and not of MDPI and/or the editor(s). MDPI and/or the editor(s) disclaim responsibility for any injury to people or property resulting from any ideas, methods, instructions or products referred to in the content.

Article

Polarization Modulation on Charge Transfer and Band Structures of GaN/MoS$_2$ Polar Heterojunctions

Feng Tian [1], Delin Kong [1], Peng Qiu [1], Heng Liu [1], Xiaoli Zhu [1], Huiyun Wei [1], Yimeng Song [1], Hong Chen [2], Xinhe Zheng [1,*] and Mingzeng Peng [1,*]

[1] Beijing Key Laboratory for Magneto-Photoelectrical Composite and Interface Science, School of Mathematics and Physics, University of Science and Technology Beijing, Beijing 100083, China
[2] Key Laboratory for Renewable Energy, Beijing Key Laboratory for New Energy Materials and Devices, Beijing National Laboratory for Condensed Matter Physics, Institute of Physics, Chinese Academy of Sciences, Beijing 100190, China
* Correspondence: xinhezheng@ustb.edu.cn (X.Z.); mzpeng@ustb.edu.cn (M.P.)

Abstract: As important third-generation semiconductors, wurtzite III nitrides have strong spontaneous and piezoelectric polarization effects. They can be used to construct multifunctional polar heterojunctions or quantum structures with other emerging two-dimensional (2D) semiconductors. Here, we investigate the polarization effect of GaN on the interfacial charge transfer and electronic properties of GaN/MoS$_2$ polar heterojunctions by first-principles calculations. From the binding energy, the N-polarity GaN/MoS$_2$ heterojunctions show stronger structural stability than the Ga-polarity counterparts. Both the Ga-polarity and N-polarity GaN/MoS$_2$ polar heterojunctions have type-II energy band alignments, but with opposite directions of both the built-in electric field and interfacial charge transfer. In addition, their heterostructure types can be effectively modulated by applying in-plane biaxial strains on GaN/MoS$_2$ polar heterojunctions, which can undergo energy band transitions from type II to type I. As a result, it provides a feasible solution for the structural design and integrated applications of hybrid 3D/2D polar heterojunctions in advanced electronics and optoelectronics.

Keywords: GaN/MoS$_2$; polarization modulation; charge transfer; band alignment; 3D/2D polar heterojunctions

1. Introduction

Due to the electronegativity difference of atoms, the positive and negative charges are separated in the wurtzite GaN and nitride semiconductors, which have extensive applications in next-generation RF power amplifiers, power switches, light-emitting diodes, lasers, solar cells, photodetectors, and integrated circuits [1–8]. GaN has two distinct polarities called Ga-polarity and N-polarity along the [0001] and [000-1] directions, respectively. The lack of inversion symmetry in GaN gives rise to a large spontaneous polarization charge of ~0.029 C/m^2 and a polarization electric field of ~10^7 V/cm. The surface polarity of GaN dramatically influences surface energies, growth modes, susceptibilities to chemicals, electric charge dynamics, and so on. Its strong polarization greatly influences the GaN-based electronic and optoelectronic performances by energy band tilt and charge separation or accumulation at the polar heterointerfaces, such as AlGaN/GaN, InGaN/GaN and metal-polar semiconductor Schottky interfaces.

With regard to the developing trend of 3D-to-2D miniaturization and multifunctional integration, 2D semiconductors beyond graphene have become a research hotspot due to their unique electrical, photonic, and mechanical performances [9]. There is a large number of 2D semiconductor materials widely used in field-effect transistors and optoelectronic devices, such as transition metal dichalcogenides (TMDs) [10,11], black phosphorus (BP) [12], silicene [13], and III-VI layered materials [14]. Among them, 2D TMDs show tunable bandgaps from 1.1 eV (MoTe$_2$) to 2.0 eV (WS$_2$) for highly visible and near-infrared optoelectronic response [15,16]. Meanwhile, 2D TMDs have excellent flexibility and high carrier mobility, enabling ultrathin and lightweight, flexible electronic applications [17].

Currently, 2D semiconductors are used as potential building blocks to form low-dimensional semiconductor heterostructures for improving electronic and optical performances [18–22], such as MoS$_2$/graphene [23], MoS$_2$/WS$_2$ [24], MoS$_2$/WSe$_2$ [25], and α-In$_2$Se$_3$/MoS$_2$ [26]. In comparison with 2D/2D architecture, mixed-dimensional 3D/2D heterostructures have greater application potential in integrating with conventional semiconductors, such as Si, GaAs, GaN, etc. In order to obtain the full utilization of the broad optical spectrum, the integration of 2D TMDs with 3D wide-bandgap GaN is a feasible solution, such as MoS$_2$/GaN, MoSe$_2$/GaN, etc. [27,28]. Desai et al. grew a homogeneous MoS$_2$ layer on a GaN substrate to produce a MoS$_2$/GaN vertical heterojunction with a type-II energy band arrangement, and showed excellent electrical rectifying characteristic and photovoltaic responsivity under monochromatic illumination [29]. Yang et al. prepared triangular MoS$_2$ monolayers on GaN substrates by CVD, which exhibited an indirect bandgap of MoS$_2$/GaN with broadband optical absorption [30]. The type-II band alignment of MoS$_2$/GaN helps to effectively separate the photogenerated electrons and holes, which demonstrated unique advantages for broadband photodetectors and high-efficiency solar cells [31–33]. Theoretically, by first-principles calculations, Gao et al. reported the MoS$_2$/2D GaN van der Waals (vdW) heterostructures possess a direct gap about 1.79 eV with type-II band alignment and excellent optical absorbance of approximately ~5.5 × 10^5 cm^{-1} [34]. It was found that changing interlayer coupling and applying external electric fields can effectively and broadly engineer its band gap and optical properties. However, 2D GaN is difficult to synthesize experimentally and does not have strong out-of-plane polarity like 3D GaN for practical polarization modulation. In addition, Sung et al. performed density-generalized function theory calculations to investigate the interfacial structures between 2D-MoS$_2$ and 3D-GaN [35]. It was reported that the surface states and electrical characteristics of the 2D/3D MoS$_2$/GaN heterostructures could be controlled by the surface terminations of GaN. However, other types of band alignments are actually needed to control carrier distribution, charge transfer, and spatial confinement of 2D/3D GaN/MoS$_2$ heterostructures. To date, there is still a lack of effective means, such as GaN polarity flip and strain modulation for optical emission, electrooptic conversion, and advanced energy-harvesting applications.

In this work, we constructed two types of GaN/MoS$_2$ (I) and (II) polar heterojunctions by manipulating the Ga-polarity and N-polarity of GaN to investigate the interfacial interactions and electronic band structures between MoS$_2$ and GaN, respectively. It is found that both of them are type-II energy band-aligned, with the heterojunction bandgaps of 0.87 eV and 1.53 eV for GaN/MoS$_2$ (I) and GaN/MoS$_2$ (II), respectively. Based on the differential charge density analysis, the negative charges are transferred in the opposite directions, from GaN to MoS$_2$ in Ga-polarity GaN/MoS$_2$ (I) and from MoS$_2$ to GaN in N-polarity GaN/MoS$_2$ (II), respectively. Correspondingly, the net charge transfer amounts of 0.036 e and 0.088 e demonstrates that the interfacial interaction is stronger in GaN/MoS$_2$ (II) than that in GaN/MoS$_2$ (I), which is consistent with the lower binding energy of GaN/MoS$_2$ (II). In addition, piezoelectric polarization of GaN has been utilized by in-plane biaxial strains to modulate the energy band configurations of GaN/MoS$_2$ heterojunctions, which can undergo an energy band transition from type II to type I. Therefore, the polarization modulation of GaN/MoS$_2$ heterojunctions provides a facile strategy for interfacial charge transfer and energy band configurations in 3D/2D polar semiconductors and device applications.

2. Computation Methods

Based on the first-principles density functional theory (DFT), our computational simulations were implemented by the Vienna ab initio simulation package (VASP) [36,37]. The heterojunction structure models were constructed by Materials Studio (MS) software. The interactions between the core and valence electrons were described by using the projector-augmented wave (PAW) method [38]. The Perdew–Burke–Ernzerhof version of the generalized gradient approximation (PBE-GGA) was selected as the electron exchange–correlation interaction functional [39], which was used for structural optimization and static self-consistent calculations. The strongly constrained and appropriately normed semilocal density (SCAN) functional was utilized to calculate the semiconductor bandgaps and electronic band structures [40]. Generally, it is efficient for accurate modeling of electronic structures of layered materials in high-throughput calculations [41]. The energy cutoff for the plane–wave basis expansion was set to be 500 eV. The Brillouin zone was sampled with a 12 × 12 × 1 Monkhorst–Pack grid. The convergence criteria of electronic and geometric optimization were 10^{-6} eV and 0.01 eV/Å. The algorithm selection for electronic optimization was normal. The Brillouin zone integration was performed by the Gaussian smearing (ISMEAR = 0) method with the smearing width (SIGMA) of 0.05 eV. The DFT-D3 method was used to correct interlayer vdW interactions between the MoS$_2$ monolayer and the GaN surface [42–45]. The vacuum space of 15 Å was adopted to avoid interactions between the periodic sheets. During the 3D/2D construction of GaN/MoS$_2$ heterostructures, the dangling bonds at the bottom surface of GaN layer were passivated by H atoms.

During our calculations, the initially established GaN/MoS$_2$ heterostructures were geometrically optimized by the lattice constants and the layer spacing. The optimized structure models were then subjected to static self-consistent calculations to obtain the charge density files. Subsequently, the SCAN generalization was used to increase the K-point grids to calculate the density of states, and the K path was set to G-M-K-G for the energy band calculation.

3. Results and Discussion

3.1. Polarity Configurations on Electronic Band Structures of GaN/MoS$_2$ vdW Heterostructures

As for the geometric structures of MoS$_2$ and GaN in Figure 1a, the 2D MoS$_2$ monolayer in the order of S-Mo-S arrangement has a hexagonal crystalline structure with the optimized lattice constants of a = b = 3.18 Å, while 3D wurtzite GaN has the same hexagonal structure with the optimized lattice constants of a = b = 3.21 Å. Their lattice mismatch between MoS$_2$ and GaN is only 0.9%, which facilitates 3D/2D heteroepitaxy for building GaN/MoS$_2$ heterostructures. Considering the spontaneous polarization effect of the GaN polar semiconductor along z axis, Ga-polarity and N-polarity GaN/MoS$_2$ heterojunctions have been constructed in the I and II stacking configurations. As shown in Figure 1b, the Ga face of GaN layer is contacted with the S atoms of MoS$_2$ monolayer at the GaN/MoS$_2$ (I) heterointerface, and, correspondingly, the N face of GaN layer is contacted with the S atoms at the GaN/MoS$_2$ (II) heterointerface, respectively. The lattice constants and interlayer equilibrium distances of the optimized GaN/MoS$_2$ unit cells are similar with a = b = 3.20 Å and d = 2.35 Å in configuration I, and a = b = 3.19 Å and d = 2.35 Å in configuration II, respectively. In order to quantitatively characterize the structural stability of GaN/MoS$_2$ heterojunctions, the binding energy has been calculated as follows,

$$E_b = E_{GaN/MoS_2} - E_{MoS_2} - E_{GaN} \tag{1}$$

where E_{GaN/MoS_2}, E_{MoS_2}, and E_{GaN} are the total energies of the GaN/MoS$_2$ vdW heterostructures, monolayer MoS$_2$ and GaN, respectively. Both GaN/MoS$_2$ (I) and GaN/MoS$_2$ (II) have negative binding energies of −0.395 eV and −0.407 eV, respectively. It demonstrates that it can form stable vdW heterostructures between MoS$_2$ and GaN. Comparatively, MoS$_2$ monolayer has a stronger interfacial interaction with the N face of GaN than that with the Ga face of GaN.

Figure 2 shows the projected band structures and local density of states (LDOS) of MoS$_2$, GaN, and the GaN/MoS$_2$ (I) and (II) heterojunctions. MoS$_2$ has a direct bandgap of 1.75 eV with both the conduction band minimum (CBM) and valence band maximum (VBM) locating at K point, as shown in Figure 2a, which is consistent with the reported experimental values [46]. In Figure 2b, GaN shows a direct bandgap of 2.4 eV at Γ point calculated by the SCAN method, which is lower than the calculated 3.4 eV by the HSE06 method. According to LDOS diagrams, both the CBM and VBM of the isolated MoS$_2$ and GaN are mainly contributed by the Mo and N states, respectively.

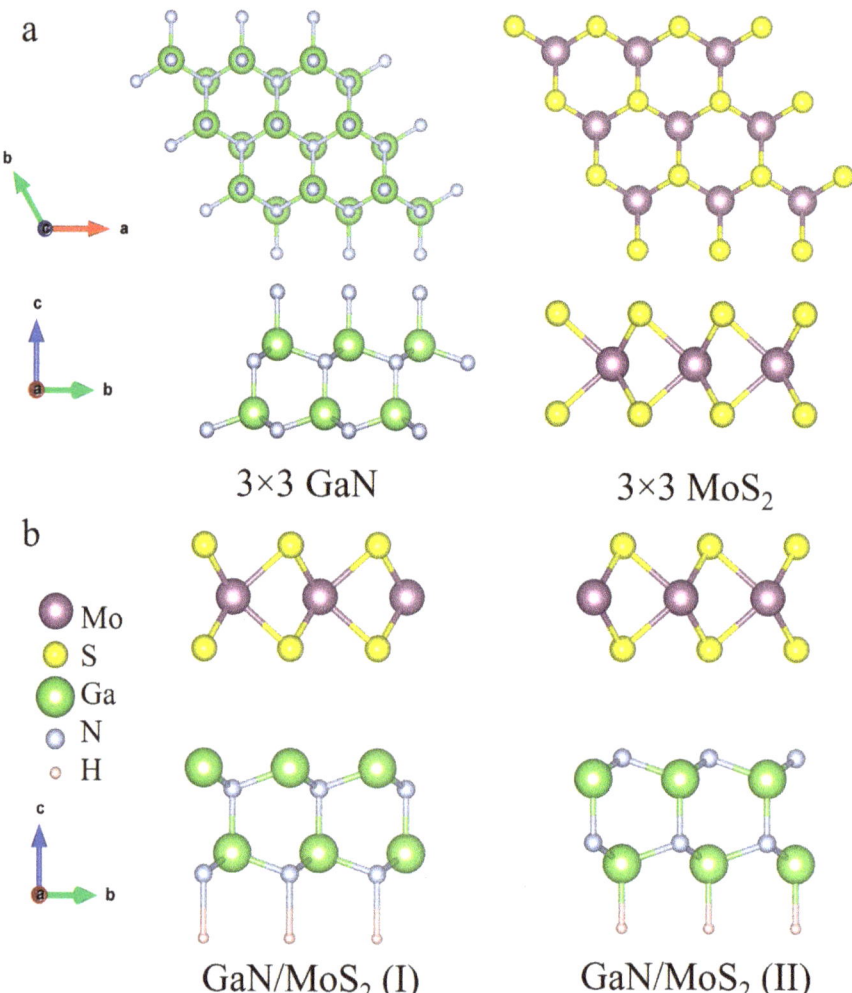

Figure 1. (a) Top and side views of 3 × 3 MoS$_2$ monolayer and 3 × 3 GaN supercells. (b) Side views of GaN/MoS$_2$ (I) and (II) heterostructures based on Ga-polarity and N-polarity surfaces, respectively.

As shown in Figure 2c,d, the energy bands of the GaN/MoS$_2$ heterojunctions are not a simple superposition of those of each GaN and MoS$_2$. The blue and red lines represent the electronic energy band weights of GaN and MoS$_2$, respectively. Due to the opposite polarity of GaN, the GaN/MoS$_2$ (I) and (II) heterojunctions have different indirect bandgaps of 0.87 eV and 1.53 eV, respectively. The CBM and VBM of GaN/MoS$_2$ (I) heterojunction is contributed by GaN and MoS$_2$, respectively. In contrast, the CBM and VBM of GaN/MoS$_2$ (II) heterojunction is contributed by MoS$_2$ and GaN, respectively. Consequently, both of them exhibit type-II band alignment, but the energy bands between MoS$_2$ and GaN can be reversed from Ga polarity to N polarity. Moreover, there exists an energy level located in the bandgap of the GaN/MoS$_2$ (II) heterojunction, as indicated by the orange arrow in

Figure 2d. In combination with the LDOS analysis, it is contributed by both of MoS$_2$ and GaN, which may arise from the interfacial hybridization of the N-polarity surface state of GaN with MoS$_2$ [35]. However, the hybridization state is not observed in the Ga-polarity GaN/MoS$_2$ (II) configuration, which is due to the surface dangling bonds of the Ga-polarity GaN bonded with the MoS$_2$.

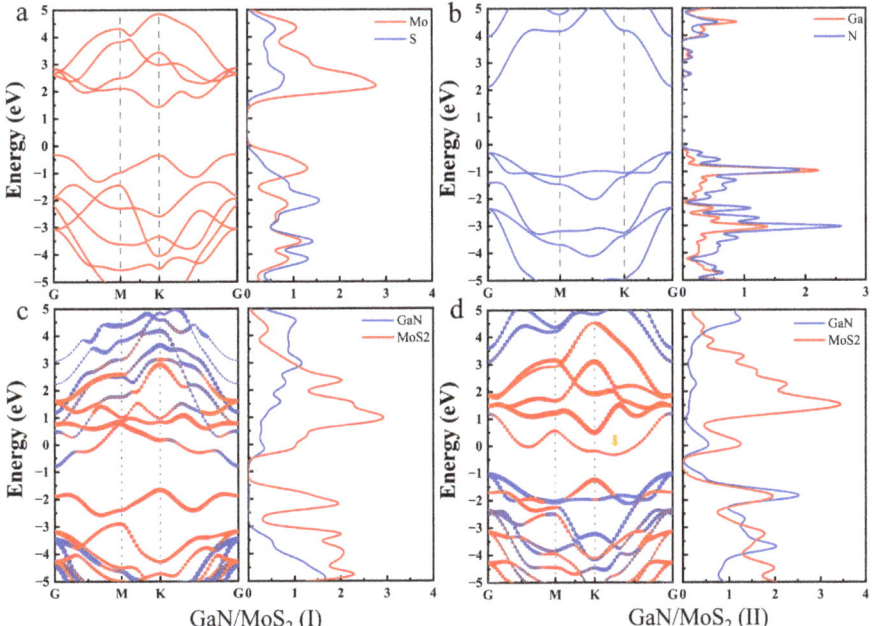

Figure 2. Band structures and local density of states of (**a**) MoS$_2$, (**b**) GaN, (**c**) GaN/MoS$_2$ (I), and (**d**) GaN/MoS$_2$ (II) heterostructures, respectively. The Fermi level is set as the zero reference.

3.2. Charge Transfer Mechanisms of GaN/MoS$_2$ vdW Heterostructures

In addition to modulating the electronic band structures, GaN polarity also has a direct impact on the charge transfer processes. Figure 3 presents the average planar electrostatic potentials of the MoS$_2$ monolayer, GaN, and GaN/MoS$_2$ (I) and (II) heterojunctions along the z direction. The black dashed line represents the energy level of a stationary electron in the vacuum (E_{VAC}) and the red dashed line represents the Fermi level of the corresponding systems (E_F). The work function (Φ) is defined as follows:

$$\Phi = E_{VAC} - E_F \qquad (2)$$

It is a critical parameter to get a further understanding on the physical mechanisms of the charge transfer at GaN/MoS$_2$ heterojunction interfaces. As shown in Figure 3a, the work function of MoS$_2$ monolayer is 5.85 eV. However, due to the spontaneous polarization induced by GaN layer, the different work functions at the left and right ends of the z-axis direction are 2.44 and 7.27 eV for GaN, 2.12 and 4.87 eV for GaN/MoS$_2$ (I), and 6.47 and 5.17 eV for GaN/MoS$_2$ (II) as shown in Figure 3b–d, respectively. There exists the electrostatic potential difference ($\Delta\Phi$) between the upper and lower surfaces of GaN, GaN/MoS$_2$ (I), and GaN/MoS$_2$ (II) heterojunctions. As presented by the interval of the black dashed lines in Figure 3b–d, the electrostatic potential differences are 4.83 eV between Ga-polarity and N-polarity surfaces of GaN, 2.75 eV for GaN/MoS$_2$ (I) and 1.3 eV for GaN/MoS$_2$ (II) configurations, respectively. It hence induces a polarization-induced electric field (E_P) according to the following equation,

$$E_P = \Delta\Phi/ed \tag{3}$$

where e and d are the elementary charge and the layer thickness of semiconducting materials or heterostructures, respectively. Correspondingly, the calculated polarization-induced electric fields are 0.68 V/Å, 0.23 V/Å, and 0.11 V/Å for GaN, GaN/MoS$_2$ (I), and GaN/MoS$_2$ (II), respectively. In particular, due to the polarity reversal of GaN/MoS$_2$ (I) and GaN/MoS$_2$ (II) configurations, their internal polarization-induced electric fields are in the opposite directions. In addition, the electrostatic potential differences of both configurations are lower than the difference of the isolated GaN. It demonstrates that the polarization strength in GaN has been attenuated by an interfacial depolarization effect. As a result, GaN polarity plays a great role on both the interfacial potential and internal electric field of GaN-based polar heterostructures.

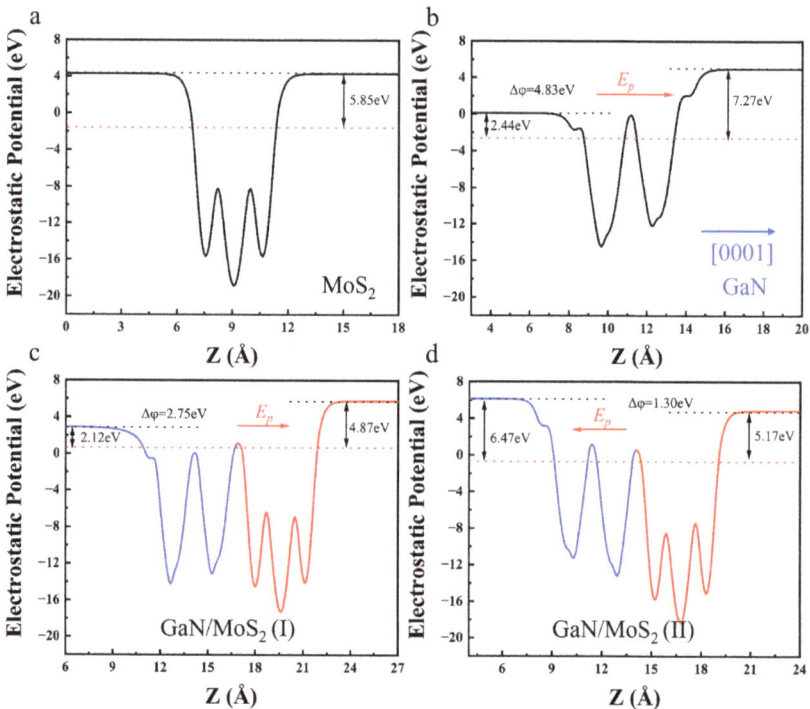

Figure 3. Planar average electrostatic potentials of (**a**) MoS$_2$ monolayer, (**b**) GaN, (**c**) GaN/MoS$_2$ (I), and (**d**) GaN/MoS$_2$ (II) heterostructures along z-direction, respectively. In (**c**) and (**d**), the blue and red parts of the electrostatic potential curves correspond to GaN and MoS$_2$, respectively. The red and blue arrows indicate the directions of the polarization-induced electric fields and [0001] polarity axis, respectively. The vacuum levels and Fermi levels are marked with black and red dashed lines, respectively.

Based on the above results of band structures and electrostatic potentials, Figure 4a,b shows the band alignment schematics of GaN/MoS$_2$ heterojunctions before contact and after contact in I and II configurations, respectively. The vacuum energy level is set to be zero as a reference. Once the GaN/MoS$_2$ heterostructures are formed, charge transfer will happen at the interface of GaN and MoS$_2$, thus achieving the same Fermi energy levels. Correspondingly, it also leads to the formation of a built-in electric field at the GaN/MoS$_2$ heterojunctions, which could hinder the further diffusion of electrons and holes, and eventually reach equilibrium. As compared with MoS$_2$, the energy bands of GaN shift downward and upward in the contacted GaN/MoS$_2$ (I) and GaN/MoS$_2$ (II) configurations in Figure 4a,b, respectively. Both the CBM and VBM of GaN are lower than those of MoS$_2$ in the contacted GaN/MoS$_2$ (I) and vice versa in the contacted GaN/MoS$_2$ (II). As a result, these two configurations have type-II band alignments, which could boost the spatial charge separation and reduce electron-hole recombination efficiency. Due to their opposite directions in terms of charge transfer and built-in electric field, the type and number of charge separation between GaN and MoS$_2$ can be directly controlled by adjusting the intrinsic polarization of GaN.

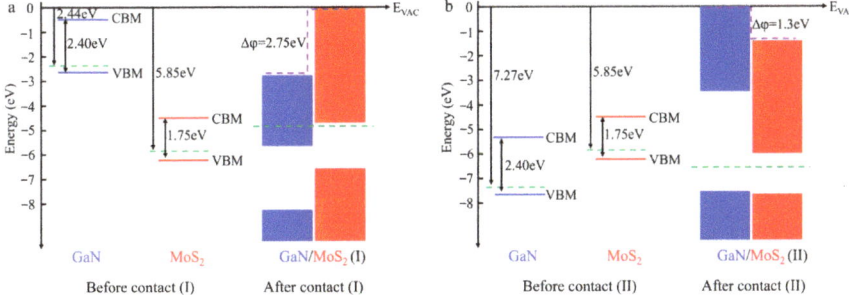

Figure 4. Schematic diagrams of energy band configurations of (**a**) GaN/MoS$_2$ (I) and (**b**) GaN/MoS$_2$ (II) heterojunctions before contact and after contact, respectively. The blue and red parts are GaN and MoS$_2$, respectively. The Fermi energy levels are indicated by green dashed lines.

In order to quantitatively analyze the interfacial charge transfer between GaN and MoS$_2$ during the formation of heterojunctions, the charge density difference ($\Delta\rho$) is calculated according to the following equation,

$$\Delta\rho = \rho(\text{GaN/MoS}_2) - \rho(\text{GaN}) - \rho(\text{MoS}_2) \tag{4}$$

where $\rho(\text{GaN/MoS}_2)$, $\rho(\text{GaN})$ and $\rho(\text{MoS}_2)$ denote the charge density of the GaN/MoS$_2$, the freestanding GaN and MoS$_2$ monolayer, respectively. Figure 5 shows the plane-averaged charge density differences of GaN/MoS$_2$ (I) and GaN/MoS$_2$ (II) along the z direction. The blue and red colors indicate the charge accumulation and depletion, respectively. In Figure 5c,d the red dashed lines represent the GaN/MoS$_2$ interface boundaries. GaN and MoS$_2$ correspond to the left and right sides of the interface boundaries, respectively. As shown in Figure 5a,c, the charges are depleted in GaN and accumulated in MoS$_2$ at the interface of GaN/MoS$_2$ (I). However, the charge depletion and accumulation are the opposite at the interface of GaN/MoS$_2$ (II) in Figure 5b,d. These results on charge transfer are consistent with their band alignments in Figure 4. In addition, the net charge transfer amounts can be evaluated by the difference between the charge accumulation and depletion at the heterointerface. They are 0.036 e and 0.088 e in GaN/MoS$_2$ (I) and (II) configurations, respectively. This demonstrates that the interfacial interaction is stronger in GaN/MoS$_2$ (II) than that in GaN/MoS$_2$ (I), which is consistent with the lower binding energy of GaN/MoS$_2$ (II).

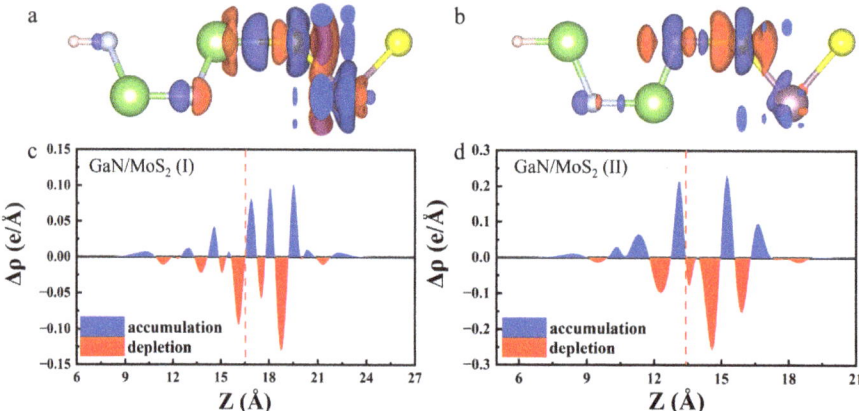

Figure 5. Plane-averaged charge density differences of (**a**) and (**c**) GaN/MoS$_2$ (I), (**b**) and (**d**) GaN/MoS$_2$ (II) heterostructures in 3D spatial distributions and along the z−direction, respectively. In (**c**) and (**d**), the red dashed lines represent the GaN/MoS$_2$ interface boundaries. GaN and MoS$_2$ correspond to the left and right sides of the interface boundaries, respectively.

3.3. Strain Modulation on the Electronic Band Structures of GaN/MoS$_2$ vdW Heterostructures

Strain engineering is of great significance to further tailor the electronic structures and band alignments of 2D polar semiconductors and heterostructures. It can be used to modulate the polarization strength based on the piezoelectric polarization effect. Here, the in-plane biaxial strain has been applied to investigate the energy bands of GaN/MoS$_2$ polar heterojunctions. The in-plane biaxial strain (ε) is defined as

$$\varepsilon = (a - a_0)/a_0 \times 100\% \tag{5}$$

where a and a_0 are the lattice constants with and without strain, respectively. Figure 6 shows the projected energy bands of the GaN/MoS$_2$ (I) and GaN/MoS$_2$ (II) polar heterojunctions by tuning the in-plane biaxial strains from −10% (compressive) to 10% (tensile). The blue and red colors indicate the contributions of GaN and MoS$_2$, respectively. For the strained GaN/MoS$_2$ (I) polar heterojunction in Figure 6a, MoS$_2$ exhibits indirect bandgaps, with VBM at K point and CBM between K and G points under compressive strains, in contrast with CBM at K point and VBM between K and G points under tensile strains. Meanwhile, GaN keeps direct band gaps with both VBM and CBM located at G point under strains from −10% to 10%. When changing from tensile strains to compressive strains, the energy bands of GaN shift significantly upward, and its band gap values decrease gradually. For the strained GaN/MoS$_2$ (II) polar heterojunction in Figure 6, MoS$_2$ shows a direct-to-indirect band gap transition from compressive strains (−6%, −10%) to tensile strains (6%, 10%), while GaN remains a direct band gap. In contrast to that of GaN/MoS$_2$ (I), a new hybridized state appeared in the band gap of GaN/MoS$_2$ (II) as indicated by the orange arrow in Figure 6b. It is due to the electron orbital interactions between surface N atoms of N-polarity GaN and MoS$_2$. When applying the biaxial tensile strains, the hybridized state gradually overlaps with the downshifted conduction bands of MoS$_2$.

In order to further clarify the energy band alignment relationship of the GaN/MoS$_2$ heterojunctions under different strains, Figure 7 presents the CBMs and VBMs of MoS$_2$ at K point and GaN at G point as a function of biaxial strains. All the CBM and VBM values of both MoS$_2$ and GaN decrease under tensile strains and increase under compressive strains in GaN/MoS$_2$ (I) and (II) configurations. In addition, the energy band alignments between MoS$_2$ and GaN can be transformed from type II to type I as applying biaxial tensile strains. The controllable band transformation is favorable for charge separation, recombination, redistribution, and fast transfer processes in 3D/2D polar heterointerfaces. Comparatively, the required tensile strain for the band transformation is less in GaN/MoS$_2$ (II) than that in GaN/MoS$_2$ (I). As a result, the biaxial strains not only modulate the electronic structures of each GaN and MoS$_2$ layer, but also play a significant role on the interfacial band hybridization and band alignment relationship of GaN/MoS$_2$ polar heterojunction.

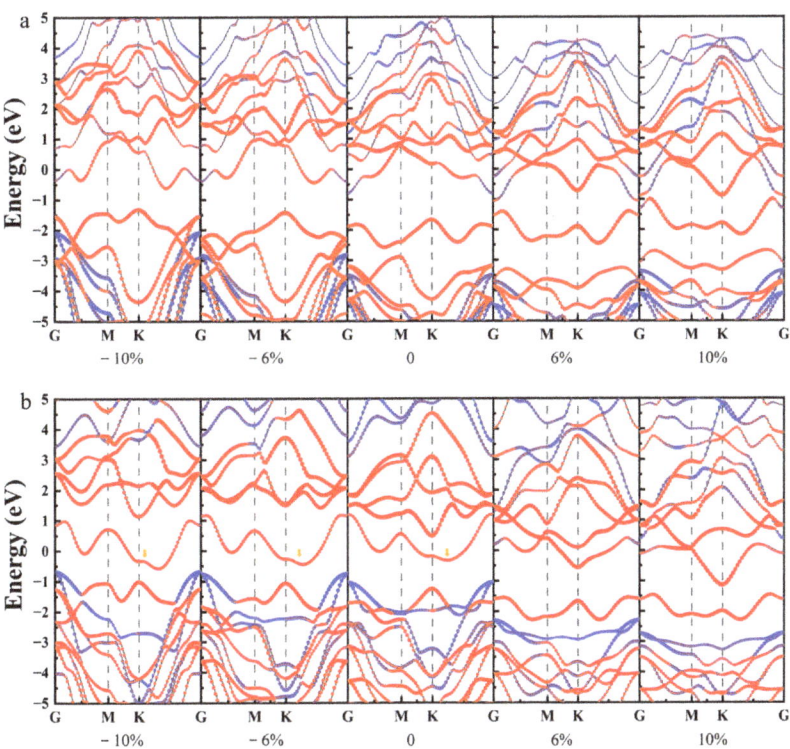

Figure 6. Projected band structures of (**a**) GaN/MoS$_2$ (I) and (**b**) GaN/MoS$_2$ (II) heterostructures under biaxial strains of −10%, −6%, 0%, 6%, and 10%, in which the red and blue colors indicates the contributions of MoS$_2$ and GaN, respectively.

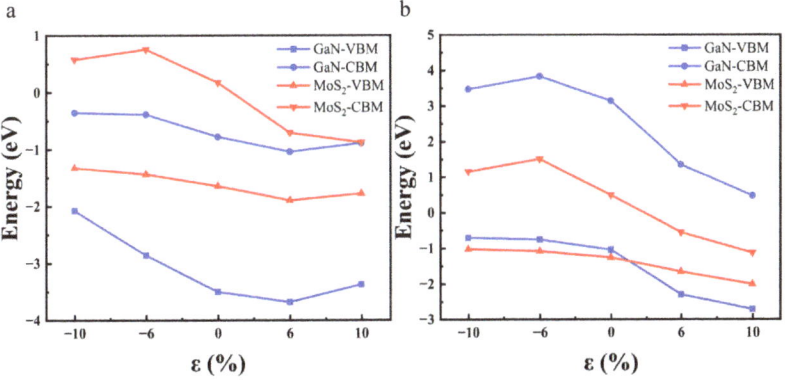

Figure 7. Band alignments of (**a**) GaN/MoS$_2$ (I) and (**b**) GaN/MoS$_2$ (II) heterostructures under biaxial strains of −10%, −6%, 0%, 6%, and 10%, respectively.

4. Conclusions

In summary, we have constructed 3D/2D GaN/MoS$_2$ polar heterojunctions to investigate the interfacial interactions and electronic band structures by polarization modulation. It demonstrates that the interfacial interaction is stronger in GaN/MoS$_2$ (II) than that in GaN/MoS$_2$ (I), as indicated by both the binding energy and net charge transfer amount. Both GaN/MoS$_2$ (I) and GaN/MoS$_2$ (II) have type-II band alignments, which could boost the spatial charge separation and reduce electron-hole recombination efficiency. By flipping the polarization directions of GaN, the type and number of transferred charges between GaN and MoS$_2$ can be directly controlled by the polarization-induced electric field. In addition, the energy band alignments between MoS$_2$ and GaN can be transformed from type II to type I as applying biaxial tensile strains. Thus, the polarization modulation is an effective way by which to control carrier distribution and spatial confinement for 3D/2D polar semiconductors and device applications.

Author Contributions: F.T., data curation, formal analysis, investigation, methodology, validation, writing—original draft; D.K., data curation, formal analysis, investigation, methodology, validation, writing—original draft. P.Q., investigation, methodology, validation. H.L., investigation, methodology, validation. X.Z. (Xiaoli Zhu), investigation, methodology, validation. H.W., investigation, methodology, validation; Y.S., investigation, methodology, validation. H.C., validation, resources, supervision; X.Z. (Xinhe Zheng), funding acquisition, resources, supervision; M.P., conceptualization, funding acquisition, project administration, resources, supervision, writing—review and editing. All authors have read and agreed to the published version of the manuscript.

Funding: This research was financially supported by the Beijing Natural Science Foundation (grant numbers 4222071), National Key R&D Program of China (grant number 2018YFA0703700), Fundamental Research Funds for the Central Universities (grant numbers FRF-BR-20-02A, FRF-TP-19-044A2, FRF-TP-20-016A2), National Natural Science Foundation of China (grant numbers 51402064), Student Research Training Project (grant number 202021010), Youth Innovation Promotion Association of Chinese Academy of Sciences (grant number 2015387).

Data Availability Statement: The data that support the findings of this study are available from the corresponding author upon reasonable request.

Conflicts of Interest: The authors declare no conflict of interest.

References

1. Mishra, U.K.; Shen, L.; Kazior, T.E.; Wu, Y.F. GaN-based RF power devices and amplifiers. *Proc. IEEE* **2008**, *96*, 287–305. [CrossRef]
2. Chen, K.J.; Häberlen, O.; Lidow, A.; Tsai, C.L.; Ueda, T.; Uemoto, Y.; Wu, Y. GaN-on-Si power technology: Devices and applications. *IEEE Trans. Electron Devices* **2017**, *64*, 779–795. [CrossRef]
3. Khan, A.; Balakrishnan, K.; Katona, T. Ultraviolet light-emitting diodes based on group three nitrides. *Nat. Photonics* **2008**, *2*, 77–84. [CrossRef]
4. Lee, K.; Wu, Z.; Chen, Z.; Ren, F.; Pearton, S.J.; Rinzler, A.G. Single wall carbon nanotubes for p-type ohmic contacts to GaN light-emitting diodes. *Nano Lett.* **2004**, *4*, 911–914. [CrossRef]
5. Choi, H.J.; Johnson, J.C.; He, R.; Lee, S.K.; Kim, F.; Pauzauskie, P.; Goldberger, J.; Saykally, R.J.; Yang, P. Self-organized GaN quantum wire UV lasers. *J. Phys. Chem. B* **2003**, *107*, 8721–8725. [CrossRef]
6. Wang, Y.; Zheng, D.; Li, L.; Zhang, Y. Enhanced efficiency of flexible GaN/perovskite solar cells based on the piezo-phototronic effect. *ACS Appl. Energy Mater.* **2018**, *1*, 3063–3069. [CrossRef]
7. Li, J.; Xi, X.; Lin, S.; Ma, Z.; Li, X.; Zhao, L. Ultrahigh sensitivity graphene/nanoporous GaN ultraviolet photodetectors. *ACS Appl. Mater. Interfaces* **2020**, *12*, 11965–11971. [CrossRef]
8. Tchernycheva, M.; Messanvi, A.; de Luna Bugallo, A.; Jacopin, G.; Lavenus, P.; Rigutti, L.; Zhang, H.; Halioua, Y.; Julien, F.H.; Eymery, J.; et al. Integrated photonic platform based on InGaN/GaN nanowire emitters and detectors. *Nano Lett.* **2014**, *14*, 3515–3520. [CrossRef]
9. Gupta, A.; Sakthivel, T.; Seal, S. Recent development in 2D materials beyond graphene. *Prog. Mater. Sci.* **2015**, *73*, 44–126. [CrossRef]
10. Zhang, Y.; Yao, Y.; Sendeku, M.G.; Yin, L.; Zhan, X.; Wang, F.; Wang, Z.; He, J. Recent progress in CVD growth of 2D transition metal dichalcogenides and related heterostructures. *Adv. Mater.* **2019**, *31*, 1901694. [CrossRef]
11. Jariwala, D.; Sangwan, V.K.; Lauhon, L.J.; Marks, T.J.; Hersam, M.C. Emerging device applications for semiconducting two-dimensional transition metal dichalcogenides. *ACS Nano* **2014**, *8*, 1102–1120. [CrossRef] [PubMed]

12. Li, L.; Yu, Y.; Ye, G.J.; Ge, Q.; Ou, X.; Wu, H.; Feng, D.; Chen, X.; Zhang, Y. Black phosphorus field-effect transistors. *Nat. Nanotechnol.* **2014**, *9*, 372–377. [CrossRef] [PubMed]
13. Tao, L.; Cinquanta, E.; Chiappe, D.; Grazianetti, C.; Fanciulli, M.; Dubey, M.; Molle, A.; Akinwande, D. Silicene field-effect transistors operating at room temperature. *Nat. Nanotechnol.* **2015**, *10*, 227–231. [CrossRef]
14. Yang, Z.; Hao, J. Recent progress in 2D layered III–VI semiconductors and their heterostructures for optoelectronic device applications. *Adv. Mater. Technol.* **2019**, *4*, 1900108. [CrossRef]
15. Chang, Y.H.; Zhang, W.; Zhu, Y.; Han, Y.; Pu, J.; Chang, J.K.; Hsu, W.; Huang, J.; Hsu, C.; Chiu, M.; et al. Monolayer $MoSe_2$ grown by chemical vapor deposition for fast photodetection. *ACS Nano* **2014**, *8*, 8582–8590. [CrossRef] [PubMed]
16. Xie, Y.; Zhang, B.; Wang, S.; Wang, D.; Wang, A.; Wang, Z.; Yu, H.; Zhang, H.; Chen, Y.; Zhao, M.; et al. Ultrabroadband MoS_2 photodetector with spectral response from 445 to 2717 nm. *Adv. Mater.* **2017**, *29*, 1605972. [CrossRef]
17. Malakar, P.; Thakur, M.S.H.; Nahid, S.M.; Islam, M.M. Data-Driven Machine Learning to Predict Mechanical Properties of Monolayer Transition-Metal Dichalcogenides for Applications in Flexible Electronics. *ACS Appl. Nano Mater.* **2022**, *5*, 16489–16499. [CrossRef]
18. Geim, A.K.; Grigorieva, I.V. Van der Waals heterostructures. *Nature* **2013**, *499*, 419–425. [CrossRef]
19. Coy Diaz, H.; Avila, J.; Chen, C.; Addou, R.; Asensio, M.C.; Batzill, M. Direct observation of interlayer hybridization and Dirac relativistic carriers in graphene/MoS_2 van der Waals heterostructures. *Nano Lett.* **2015**, *15*, 1135–1140. [CrossRef]
20. Novoselov, K.S.; Mishchenko, A.; Carvalho, O.A.; Castro Neto, A.H. 2D materials and van der Waals heterostructures. *Science* **2016**, *353*, aac9439. [CrossRef]
21. Pierucci, D.; Henck, H.; Avila, J.; Balan, A.; Naylor, C.H.; Patriarche, G.; Dappe, Y.J.; Silly, M.G.; Sirotti, F.; Johnson, A.T.C.; et al. Band alignment and minigaps in monolayer MoS_2-graphene van der Waals heterostructures. *Nano Lett.* **2016**, *16*, 4054–4061. [CrossRef]
22. Le, P.T.T.; Hieu, N.N.; Bui, L.M.; Phuc, H.V.; Hoi, B.D.; Amin, B.; Nguyen, C.V. Structural and electronic properties of a van der Waals heterostructure based on silicene and gallium selenide: Effect of strain and electric field. *Phys. Chem. Chem. Phys.* **2018**, *20*, 27856–27864. [CrossRef] [PubMed]
23. Bertolazzi, S.; Krasnozhon, D.; Kis, A. Nonvolatile memory cells based on MoS_2/graphene heterostructures. *ACS Nano* **2013**, *7*, 3246–3252. [CrossRef] [PubMed]
24. Tongay, S.; Fan, W.; Kang, J.; Park, J.; Koldemir, U.; Suh, J.; Narang, D.S.; Liu, K.; Ji, J.; Li, J.; et al. Tuning interlayer coupling in large-area heterostructures with CVD-grown MoS_2 and WS_2 monolayers. *Nano Lett.* **2014**, *14*, 3185–3190. [CrossRef] [PubMed]
25. Shim, G.W.; Yoo, K.; Seo, S.B.; Shin, J.; Jung, D.Y.; Kang, I.S.; Ahn, C.W.; Cho, B.J.; Choi, S.Y. Large-area single-layer $MoSe_2$ and its van der Waals heterostructures. *ACS Nano* **2014**, *8*, 6655–6662. [CrossRef]
26. Zhou, B.; Jiang, K.; Shang, L.; Zhang, J.; Li, Y.; Zhu, L.; Gong, S.J.; Hu, Z.; Chu, J. Enhanced carrier separation in ferroelectric In_2Se_3/MoS_2 van der waals heterostructure. *J. Mater. Chem. C* **2020**, *8*, 11160–11167. [CrossRef]
27. Chen, Z.; Liu, H.; Chen, X.; Chu, G.; Chu, S.; Zhang, H. Wafer-size and single-crystal $MoSe_2$ atomically thin films grown on GaN substrate for light emission and harvesting. *ACS Appl. Mater. Interfaces* **2016**, *8*, 20267–20273. [CrossRef]
28. Yang, G.; Gu, Y.; Yan, P.; Wang, J.; Xue, J.; Zhang, X.; Lu, N.; Chen, G. Chemical vapor deposition growth of vertical MoS_2 nanosheets on p-GaN nanorods for photodetector application. *ACS Appl. Mater. Interfaces* **2019**, *11*, 8453–8460. [CrossRef]
29. Desai, P.; Ranade, A.K.; Shinde, M.; Todankar, B.; Mahyavanshi, R.D.; Tanemura, M.; Kalita, G. Growth of uniform MoS_2 layers on free-standing GaN semiconductor for vertical heterojunction device application. *J. Mater. Sci. Mater. Electron.* **2020**, *31*, 2040–2048. [CrossRef]
30. Yang, G.; Ding, Y.; Lu, N.; Xie, F.; Gu, Y.; Ye, B.; Yao, Y.; Zhang, X.; Huo, X. Insights Into the Two-Dimensional MoS_2 Grown on AlGaN (GaN) Substrates by CVD Method. *IEEE Photonics J.* **2021**, *13*, 2200105. [CrossRef]
31. Cao, B.; Ma, S.; Wang, W.; Tang, X.; Wang, D.; Shen, W.; Qiu, B.; Xu, B.; Li, G. Charge Redistribution in Mg-Doped p-Type MoS_2/GaN Photodetectors. *J. Phys. Chem. C* **2022**, *126*, 18893–18899. [CrossRef]
32. Zhang, X.; Li, J.; Ma, Z.; Zhang, J.; Leng, B.; Liu, B. Design and integration of a layered MoS_2/GaN van der Waals heterostructure for wide spectral detection and enhanced photoresponse. *ACS Appl. Mater. Interfaces* **2020**, *12*, 47721–47728. [CrossRef] [PubMed]
33. Hassan, M.A.; Kim, M.W.; Johar, M.A.; Waseem, A.; Kwon, M.K.; Ryu, S.W. Transferred monolayer MoS_2 onto GaN for heterostructure photoanode: Toward stable and efficient photoelectrochemical water splitting. *Sci. Rep.* **2019**, *9*, 20141. [CrossRef] [PubMed]
34. Gao, R.; Liu, H.; Liu, H.; Yang, J.; Yang, F.; Wang, T. Two-dimensional MoS_2/GaN van der Waals heterostructures: Tunable direct band alignments and excitonic optical properties for photovoltaic applications. *J. Phys. D Appl. Phys.* **2019**, *53*, 095107. [CrossRef]
35. Sung, D.; Min, K.A.; Hong, S. Investigation of atomic and electronic properties of 2D-MoS_2/3D-GaN mixed-dimensional heterostructures. *Nanotechnology* **2019**, *30*, 404002. [CrossRef]
36. Kresse, G.; Furthmüller, J. Efficiency of ab-initio total energy calculations for metals and semiconductors using a plane-wave basis set. *Comput. Mater. Sci.* **1996**, *6*, 15–50. [CrossRef]
37. Blöchl, P.E. Projector augmented-wave method. *Phys. Rev. B* **1994**, *50*, 17953. [CrossRef]
38. Kresse, G.; Joubert, D. From ultrasoft pseudopotentials to the projector augmented-wave method. *Phys. Rev. B* **1999**, *59*, 1758. [CrossRef]
39. Perdew, J.P.; Burke, K.; Ernzerhof, M. Generalized gradient approximation made simple. *Phys. Rev. Lett.* **1996**, *77*, 3865. [CrossRef]

40. Sun, J.; Ruzsinszky, A.; Perdew, J.P. Strongly constrained and appropriately normed semilocal density functional. *Phys. Rev. Lett.* **2015**, *115*, 036402. [CrossRef]
41. Buda, I.G.; Lane, C.; Barbiellini, B.; Ruzsinszky, A.; Sun, J.; Bansil, A. Characterization of thin film materials using SCAN meta-GGA, an accurate nonempirical density functional. *Sci. Rep.* **2017**, *7*, 44766. [CrossRef] [PubMed]
42. Grimme, S. Semiempirical GGA-type density functional constructed with a long-range dispersion correction. *J. Comput. Chem.* **2006**, *27*, 1787–1799. [CrossRef] [PubMed]
43. Grimme, S.; Antony, J.; Ehrlich, S.; Krieg, H. A consistent and accurate ab initio parametrization of density functional dispersion correction (DFT-D) for the 94 elements H-Pu. *J. Chem. Phys.* **2010**, *132*, 154104. [CrossRef]
44. Grimme, S.; Ehrlich, S.; Goerigk, L. Effect of the damping function in dispersion corrected density functional theory. *J. Comput. Chem.* **2011**, *32*, 1456–1465. [CrossRef] [PubMed]
45. Kerber, T.; Sierka, M.; Sauer, J. Application of semiempirical long-range dispersion corrections to periodic systems in density functional theory. *J. Comput. Chem.* **2008**, *29*, 2088–2097. [CrossRef] [PubMed]
46. Mak, K.F.; Lee, C.; Hone, J.; Shan, J.; Heinz, T.F. Atomically thin MoS_2: A new direct-gap semiconductor. *Phys. Rev. Lett.* **2010**, *105*, 136805. [CrossRef]

Disclaimer/Publisher's Note: The statements, opinions and data contained in all publications are solely those of the individual author(s) and contributor(s) and not of MDPI and/or the editor(s). MDPI and/or the editor(s) disclaim responsibility for any injury to people or property resulting from any ideas, methods, instructions or products referred to in the content.

Article

Micro-Raman Spectroscopy Study of Vertical GaN Schottky Diode

Atse Julien Eric N'Dohi [1], Camille Sonneville [1,*], Soufiane Saidi [1], Thi Huong Ngo [2], Philippe De Mierry [2], Eric Frayssinet [2], Yvon Cordier [2], Luong Viet Phung [1], Frédéric Morancho [3], Hassan Maher [4] and Dominique Planson [1]

[1] Univ. Lyon, Université Claude Bernard Lyon 1, INSA Lyon, Ecole Centrale Lyon, CNRS, Ampère, UMR5005, 69621 Villeurbanne, France
[2] CNRS, CRHEA, Université Côte d'Azur, 06560 Valbonne, France
[3] LAAS-CNRS, Université de Toulouse, CNRS, UPS, 31031 Toulouse, France
[4] LN2-CNRS-UMI, Université de Sherbrooke, Sherbrooke, QC J1K 2R1, Canada
* Correspondence: camille.sonneville@insa-lyon.fr

Abstract: In this work, the physical and the electrical properties of vertical GaN Schottky diodes were investigated. Cathodo-luminescence (CL), micro-Raman spectroscopy, SIMS, and current-voltage (I-V) measurements were performed to better understand the effects of physical parameters, for example structural defects and doping level inhomogeneity, on the diode electrical performances. Evidence of dislocations in the diode epilayer was spotted thanks to the CL measurements. Then, using 2D mappings of the E_2^h and A_1 (LO) Raman modes, dislocations and other peculiar structural defects were observed. The I-V measurements of the diodes revealed a significant increase in the leakage current with applied reverse bias up to 200 V. The combination of physical and electrical characterization methods indicated that the electrical leakage in the reverse biased diodes seems more correlated with short range non-uniformities of the effective doping than with strain fluctuation induced by dislocations.

Keywords: cathodo-luminescence; micro-Raman spectroscopy; current-voltage I-V; GaN Schottky diodes; power electronic devices

1. Introduction

Gallium nitride (GaN) has attracted much attention for their potential applications in high voltage electronic devices due to their superior physical properties, such as a wide band gap energy, high electron mobility, large breakdown field, and high thermal conductivity [1,2]. To date, commercially available lateral GaN power devices such as high-electron-mobility transistors (HEMTs) fabricated on foreign substrates have shown good and effective electrical performance [3]. However, for most of industrial GaN-on-Silicon HEMTs, this performance is still limited to a breakdown voltage of 650 V. This limitation is attributed to the reported lattice-mismatch-induced dislocations and thermo-elastic strain resulting in a limited thickness of the buffer layer. Vertical GaN devices are reported to be more suitable for high power applications versus lateral ones [4,5]. Indeed, it has been shown that a vertical structure is the most efficient way to increase both the breakdown voltage (BV) and the current density. Moreover, vertical devices should be less sensitive to surface states contrary to HEMTs. In order to achieve these performances, critical aspects such as high structural quality of the drift region, doping control, and its homogeneity need to be addressed. This is possible through an efficient use of physical and electrical characterization approaches. Recent studies have been carried out to examine the crystalline properties and electrical behavior of vertical GaN Schottky concomitantly. Ren et al. observed that the microstructure of the GaN layers and electrical properties of Schottky Barrier Diodes (SBDs) were strongly dependent on the epitaxial growth rates. By

optimizing the growth conditions, they obtained a high structural quality GaN drift layer with high mobility [4]. Tompkins et al. have also shown that deep acceptor states associated with carbon lead to a decrease in the breakdown while increasing the specific on-resistance in the GaN SBDs grown at 100 Torr [6]. Other research groups mentioned the effect of threading dislocations and the doping concentration on the electrical behavior of GaN vertical PN diodes [7,8]. Nonetheless, a clear physical understanding of the correlation between the microstructural properties and the electrical performance of the device is still not widely established.

In this work, non-destructive physical characterizations such as cathodo-luminescence (CL) and micro-Raman spectroscopy were performed and coupled with electrical characterization (reverse and forward I–V) to assess the effects of structural and electrical defects on the electrical performance of the vertical GaN SBDs. Indeed, dislocation clusters can be highlighted by CL measurements, due to their non-radiative recombination activity, and the structural defects [9] and the n-doping concentration distribution [10] can be probed by 2D mapping Raman spectroscopy. In this study, the E_2^h peak position and width were tracked to analyze the structural defects and stress distribution and the A_1 (LO) peak position and intensity are used to determine the n carrier concentration and homogeneity. Moreover, the I-V measurements were performed to check electrical performance of the SBDs. The combination of physical and electrical characterization methods indicated that the electrical leakage in the reverse biased diodes seems more correlated with short range non-uniformities of the effective doping than with the strain fluctuation induced by the dislocations.

2. Materials and Methods

A 5 μm Si n-doped layer was grown by MOCVD (Metal Organic Chemical Vapor Deposition) method on a GaN HVPE freestanding substrate from Saint-Gobain Lumilog [11]. GaN films were grown in a close-coupled showerhead reactor. Ammonia, trimethylgallium (TMGa), and hydrogen carrier gas were used to grow the films at 1020 °C at a growth rate of 2 μm.h^{-1}. Diluted silane was added to the vapor phase in order to dope the GaN film. The effective doping concentration value amounts to $N_d - N_a = 8 \times 10^{15}$ cm^{-3} and was determined by mercury probe capacitance-voltage (C-V) technique [12]. First, 40 nm thick rectangular Ni frames were defined by photolithography, electron beam deposition, and lift off. Then, CL measurements were performed before any contact deposition. The Schottky contact (Ni/Au) and the backside ohmic contact (Ti/Al/Ni/Au) were electron beam deposited (Figure 1a). The diodes were mesa-isolated by chlorine-based reactive ion etching. Four SBDs with different diameters (200 μm, 100 μm, and 50 μm) were fabricated on each nickel frame (Figure 1b). C-V measurements performed on these Schottky diodes confirm the previously determined concentration in the epilayer [12].

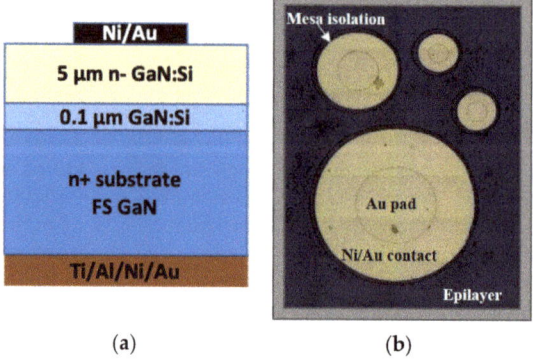

Figure 1. (a) Cross section of the vertical GaN Schottky diodes; (b) Optical microscope view of the fabricated diodes with 50 μm, 100 μm, and 200 μm diameter.

Micro-Raman spectroscopy measurements were carried out at room temperature using a confocal spectrometer (Renishaw Invia model) in back scattering geometry with a ×100 objective and a 2400 L/mm diffraction grating and a 532 nm laser excitation. The micro-Raman measurements were carried out after the complete removal of the Ni/Au Schottky contact by chemical etching. The spectral and spatial lateral resolutions were found to be around 0.1 cm^{-1} and 1 µm, respectively, and the depth resolution was between 3 and 5 µm. For the Raman measurements, 2D Raman maps were made on the epilayer of the diodes after the metallization removal. During 2D Raman measurements, cartographies of 500 × 350 µm size with a step size of 5 µm were performed on each frame containing the diodes. From these measurements, a series of E_2^h and A_1 (LO) Raman spectra were obtained and fitted using a mixed Gaussian–Lorentzian function with the Wire^{TM} Renishaw software (Figure 2). From these fitting, we extracted the values of E_2^h and A_1 (LO) peak position and intensity to create the Raman maps that were used to probe the physical properties of the diodes.

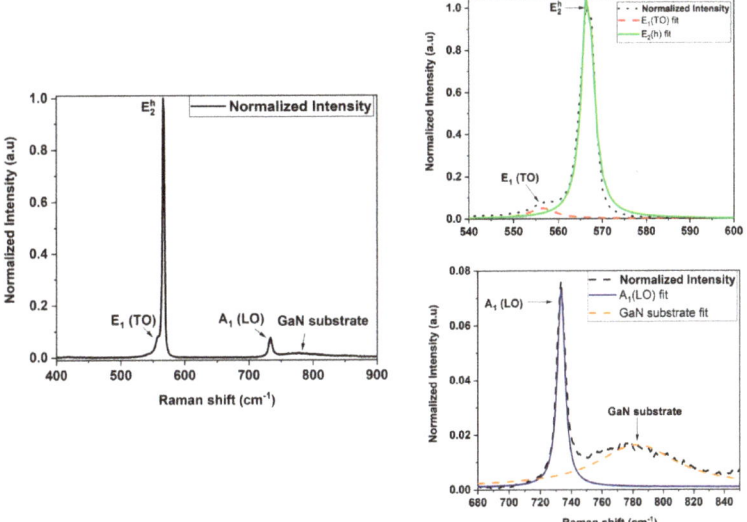

Figure 2. Raman spectrum of the diode epilayer (**left**). Right: fitted curves of E_2^h (**upper right**) and A_1 (LO) modes (**lower right**).

The CL images were recorded at room temperature using an acceleration voltage of 10 kV and a magnification of ×500 or ×1000. By inspection of 15 CL images, the dislocation density was found to be 5–10 × 10^6 cm^{-2} in 187 × 500 µm^2 areas [12]. This technique allows us to inspect the arrangement of dislocations in the GaN epilayers before the fabrication of metal contacts.

3. Results

3.1. SIMS Characterization

The SIMS measurements were conducted to examine the amount of impurities in the drift layer of the diodes and previously described in reference [13]. Figure 3 illustrates the SIMS depth profile for the Si-doped GaN film. The presence of background impurities such as carbon and oxygen can clearly be seen in the drift layer. The SIMS profile shows a very high Si concentration from the surface to about 100 nm dept. This has been previously reported in the following references [13–15] and can be explained by surface dopant atom contamination (see reference [13] for more details). After a 100 nm depth, the SIMS profile shows a uniform Si doping density (~2.5 × 10^{16} cm^{-3}) corresponding to the *n*-doped GaN epilayer grown by MOCVD. Beyond the depth of 4.2 µm, a high concentration of Si is observed corresponding to a 100 nm thin layer of highly doped Si GaN. This high-doped layer is regrown during the MOCVD process on top of the freestanding GaN substrate to avoid interface contaminations such as carbon impurities that limit the series resistance. For this reason, high concentration of Si dopants is present in this region located beyond the depth of 4.2 µm. Finally, the 4.5 µm depth corresponds to the freestanding GaN substrate, as the substrate is doped with oxygen and the HVPE process limits the amount of carbon. Overall, the SIMS results reveal that the silicon concentration is almost uniform (~2.5 × 10^{16} cm^{-3}) in the GaN active layer. The SIMS results highlighted that the presence of carbon and oxygen impurities is not negligible. From the surface down to a 4 µm depth in the epilayer, the O and C concentrations are constant. Oxygen is a shallow donor and is present in the films due to the contamination from ammonia source [16]. On the other hand, carbon is a deep acceptor and its presence in the film is due to the decomposition of TMGa methyl groups during MOCVD process [17]. Hence, some compensation effects of the Si and O donors by the C acceptors may occur.

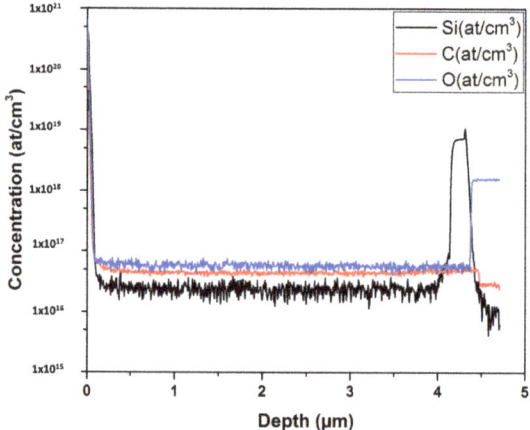

Figure 3. SIMS profiles measurement from the studied GaN vertical structure. The black curve highlights the presence of Si dopants (main dopant). The red and blue curves represent the main background impurities in the sample (carbon and oxygen respectively).

3.2. Study of the E_2^h Peak Behavior

In epitaxial GaN, the stress is biaxial and the E_2^h mode is reported to be sensitive to the biaxial stress [18] and useful to probe crystalline quality as well; its non-polar nature makes all atoms vibrate in the x-y plane. With this mode, any effect on the atomic bonds in the lattice can be effectively sensed. Therefore, a shift of this mode indicates the level or the type of stress in the epilayer and can be accounted by Equation (1): [18]

$$\Delta \omega = K \sigma_{xx=yy} \qquad (1)$$

where $\Delta \omega$ is the shift of phonon line (cm^{-1}), $\sigma_{xx=yy}$ (GPa) is the biaxial stress and K (cm^{-1}/GPa) is the pressure coefficient or the stress coefficient. That expression is useful to quantify the residual stress in the samples as long as the pressure coefficient is known. We did not evaluate the stress due to the scattered value of K in the literature [18–20]. Here, by investigating the E_2^h 2D Raman maps obtained as described earlier in the experimental method on each Ni frame, the stress and the crystal quality of the diodes drift layer were assessed. Actually, the wafer containing the homo-epi-structure was cut into two parts. The first sample contains frames 1 and 2 while the second sample contains frames 3 and 4. Figure 4 shows the E_2^h Raman maps and the CL maps (respectively in black and white) performed on the four different frames on the GaN wafer.

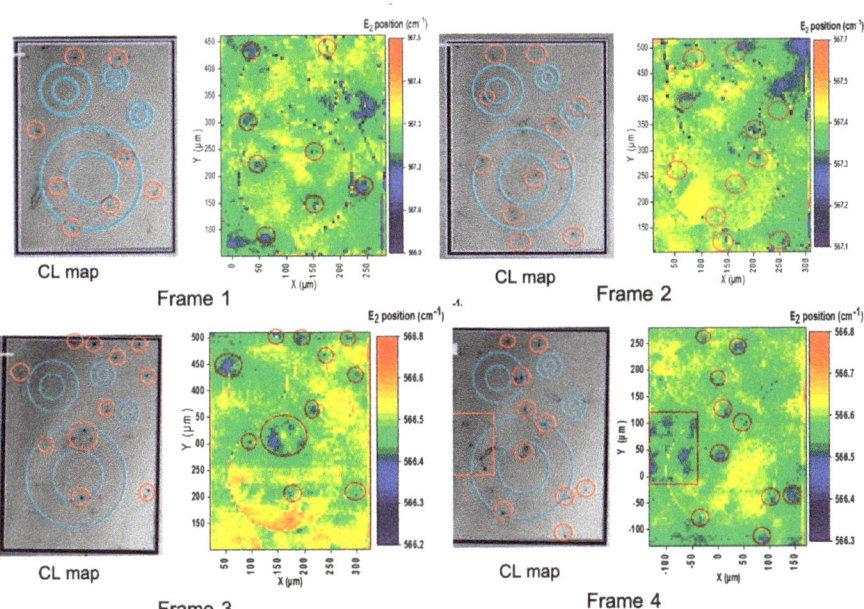

Figure 4. 2D Raman maps of E_2^h peak position (right) and their corresponding cathodo-luminescence images (in black and white, left) performed on the four frames. The red circles and rectangle are ascribed to dislocations clusters. The blue circles correspond to the areas where Schottky diodes were fabricated afterwards.

Table 1 summarizes the supporting quantification results extracted from the E_2^h position and width maps obtained by fittings. We clearly see in the Table 1 the difference in the E_2^h position mean value. Frames 1 and 2 have the E_2^h position mean value between 567.3 and 567.4 cm^1 range while frames 3 and 4 show a lower shift between 566.5 and 566.6 cm^{-1}. That difference of about 0.8 cm^{-1} may be due to the long range inhomogeneities of the original sample or to the fact that frames 3 and 4 have undergone many physical experiments such as SEM, I-V, Raman, and temperature measurements: It may have been structurally deteriorated by these operations. Therefore, we did not examine the A_1(LO) Raman map of these samples. According to Equation (1) and assuming that the mean value corresponds to the strain relaxed state, we can see on the E_2^h position maps a shift below (toward the blue zone) or above (toward the red zone) the mean value, which indicates a tensile or compressive stress, respectively. From the comparison between the E_2^h position maps and the CL image, we identified the presence of the dislocation clusters. Indeed, the observed dislocations in the CL image match well zone by zone with a shift of the E_2^h peak to a lower position in the Raman map (Figure 4). These results established a good correlation between the Raman spectroscopy and CL and therefore proves the efficiency of the micro-Raman spectroscopy as a non-destructive tool to highlight the presence of dislocations clusters as reported previously by Kokubo et al. [21]. Generally, three types of threading dislocations exist: the edge dislocations (TEDs), the screw dislocations (TSDs), and the mixed dislocations (edge and screw) (TMDs). In references [21,22] it has been reported that the TSDs do not affect the E_2^h peak shift because the shear strain has less influence on the E_2^h shift. Thus, the screw type dislocations are not detectable by means of the Raman peak shift. However, it is still challenging to distinguish the TED from the TMD types when they are both present in a GaN sample. In these previous studies, the dislocations appear as zones affected by a shift of the E_2^h peak position to a higher position close to a zone with a shift of the E_2^h peak position to a lower position. In our study, it seems that the observed dislocation clusters appear only in blue on the Raman E_2^h position map and then correspond to an area with a lower stress. This reveals that the dislocation clusters may induce a tensile stress in the drift layer of the studied diodes. In addition, it is possible that either the step used for our mapping is too large to observe the compressive stress or that the dislocations observed here are dislocation clusters locally decreasing the stress in a similar way as the dislocations associated with the grain boundaries [23]. From the E_2^h width maps (not presented here, results summarized in Table 1), we see that the frames present peak widths with the same order of magnitude (3.6~ 3.7 cm^{-1}). In frames 1 and 2, the value of v = 3.6 cm^{-1} can be noticed while in frames 3 and 4 the value of v = 3.7 cm^{-1} is obtained. We note a small E_2^h width gap shift of 0.1 cm^{-1} among the frames. This gap shift is insignificant to account for an effective crystalline inhomogeneity in the frames. Hence, we can tentatively infer that the crystalline structure of the drift layer is homogeneous over the whole epi wafer. Moreover, from the reference [24], the value up to v = 3.7 cm^{-1} indicates that the Si-doped epilayer is of good crystalline quality.

Table 1. E_2^h position and width quantification result.

Frame Name	E_2^h Position (Mean Value) (cm^{-1})	E_2^h Width (Mean Value) (cm^{-1})
Frame 1	567.3 ± 0.1	3.6 ± 0.1
Frame 2	567.4 ± 0.1	3.6 ± 0.1
Frame 3	566.5 ± 0.1	3.7 ± 0.1
Frame 4	566.6 ± 0.1	3.7 ± 0.1

3.3. Study of the A_1 (LO) Peak Behavior

The A_1(LO) phonon mode is used to measure the spatial distribution of the free-carrier concentration and carrier mobility of polar semiconductors such as GaN due to the interaction between the longitudinal optical (LO) phonon modes and the collective oscillation of free carriers (plasmons) that forms the LO Phonon-Plasmon coupled (LOPC) mode [25–27]. The shift in the A_1 (LO) position is known to reflect changes in the plasma frequency and enables to estimate the free-carrier density and/or mobility. The evaluation and the optimization of the carrier concentration of the GaN-based device drift layer is important for their electrical performance. By tracking the A_1 (LO) position shift through a Raman mapping, we determined the effective carrier concentration of the diodes. It has been reported that for n-doped GaN layers with a n concentration below 10^{17} cm^{-3}, the A_1 (LO) position shift is a linear function of the carrier concentration through Equation (2) [28]:

$$\omega_1 = 1.410^{-17} n + \omega_0 \qquad (2)$$

where n is the n-carrier concentration, ω_1(A_1(LO)) is the Raman shift in cm^{-1} and ω_0 is an offset value of 733.3 cm^{-1} deduced from the plot. This means that any A_1 (LO) position shift corresponds to a specific carrier concentration. As mentioned above, only frames 1 and 2 were considered in this part. From the A_1(LO) 2D position maps as displayed in Figure 5, this implies that the n-carrier mean concentration does not significantly vary across the region containing these frames. The doping is therefore rather homogeneous all over the region. This value corresponds to nearly $n = 7 \times 10^{15}$ cm^{-3} (Table 2) as an estimated carrier concentration using Equation (2). We clearly realize that we find a similar concentration as determined by the capacitance-voltage (C-V) mercury probe method ($n = 8 \times 10^{15}$ cm^{-3}). Therefore, we observe a good correlation between the Raman spectroscopy and the C-V method in terms of the carrier concentration evaluation.

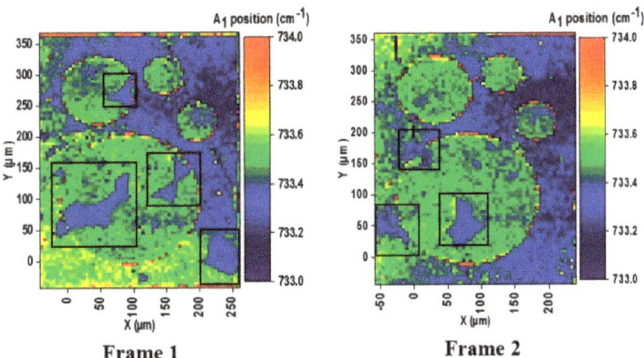

Figure 5. 2D Raman maps of $A_1(LO)$ peak position of frames 1 and 2. The black squares are ascribed to peculiar patches.

Table 2. $A_1(LO)$ position and n-carrier concentration quantification result.

Frame Name	$A_1(LO)$ Position (Mean Value) (cm^{-1})	$A_1(LO)$ Width (Mean Value) (cm^{-1})
Frame 1	733.4 ± 0.1	7.1 × 10^{15} ± 10%
Frame 2	733.4 ± 0.1	7.1 × 10^{15} ± 10%

Moreover, when considering both the $A_1(LO)$ intensity and the position maps of frame 1 and 2 (Figures 5 and 6), we observe peculiar patches (highlighted in black squares) in the active area of the 200 µm diodes. These patches did not appear in the E_2^h maps as shown in Figure 3. Therefore, as no local strain changes are noticed, these patches cannot be ascribed to dislocations. Rather, these patches may stand for the inhomogeneous incorporation of impurities (such as carbon) during the MOCVD growth process. In these peculiar patches, the $A_1(LO)$ intensity increases (red part) while the Raman shift position decreases (blue zone) as shown in Figures 5 and 6. As a result, they seem to affect the doping and its homogeneity in the diodes by apparently locally decreasing either the n-free-carrier concentration or their mobility. According to the SIMS results, they may correspond to local changes in the incorporation of silicon or background impurities such as oxygen (O) or carbon (C) or to an agglomeration of them since they all exist together in the probed epilayer thickness (5 µm). Further analyses such as EBIC [29,30] are needed to determine the exact nature of the defects that create this kind of inhomogeneity.

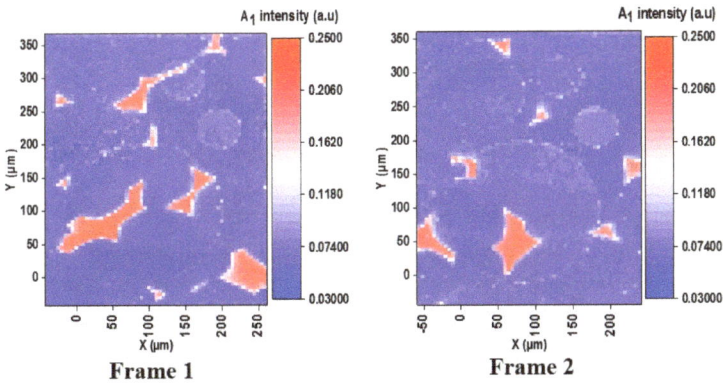

Figure 6. 2D Raman maps of $A_1(LO)$ peak intensity of frames 1 and 2.

3.4. Electrical Characterization

The electrical behavior of the vertical SBDs was investigated through the I-V forward measurements. Figure 7 shows the result of the forward I-V characterization obtained on the four frames, containing four diodes each. The 50 μm (high) and 50 μm (low) refer to the respective location of the 50 μm diameter diodes in each Ni frame. We notice that the current density in the diodes have all the same exponential dependence when the voltage increases from 0.1 V to 0.4 V. It also shows that they are dominated by the same forward current conduction mechanism independently of the diode size. The exponential part of the I–V curves can be explained by the thermionic emission model. In this region, the current density J can be expressed with the following equation [31]:

$$J = A^* T^2 \exp\left(-\frac{q\phi_B}{kT}\right) \exp\left(\frac{qV}{nkT}\right) \qquad (3)$$

where A^* is the GaN Richardson's constant for GaN, T is the temperature, ϕ_B is the Schottky barrier height (SBH), k is the Boltzmann constant, V is the applied bias voltage, q is the elementary charge, n is the ideality factor. When the forward bias is beyond 0.4 V, the increase in the current density is limited by the series resistance, and Equation (3) should be modified. From Equation (3), applied to the linear region of the log(J)-V curve, we deduce the ideality factor from the intercept and the barrier height from the slope. Table 3 summarizes the values obtained for n and Φ_B. All the diodes show the typical Schottky behavior with the ideality factor close to the unity and varying from 1.01 up to 1.19, independently of the diode size. The barrier heights are between 0.74 V and 0.91 V and seem to increase with the diode size.

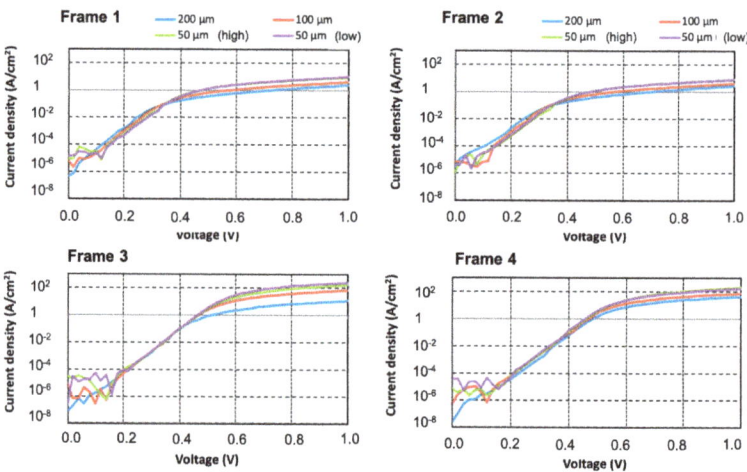

Figure 7. Forward I-V characteristics for diodes from the frames (1, 2, 3 and 4): blue (200-μm diode), red (100-μm diode), green (50-μm high), purple (50-μm low).

The diode electrical behavior was investigated through the I–V reverse measurements down to −200 V. Figure 8 shows the result of the reverse I–V characterization. All the studied diodes on the wafer display nearly the same reverse characteristics with a relatively high leakage current density ranging between 10^{-3} and 10^{-1} A.cm^{-2} at −200 V. The third column of Table 3 summarizes the reverse voltage (V_r), at which a leakage current density of 0.5×10^{-2} A.cm^{-2} is reached. This voltage varies between −70 V and −185 V. The lower V_r is obtained for the diodes of frames 1 and 2 and the higher V_r for the diodes of frames 3 and 4. However, V_r seems to be independent of the diode size. Then we suspect that the observed dispersion between frames may be caused by the presence of some structural defects (such as dislocations) or electrical defects in the drift layer. These defects may actively play a major role in the performance of the diodes by somehow influencing the net carrier distribution across the drift layer.

Table 3. Ideality factor (n) and Schottky barrier height (ϕ_B) deduced from the I-V forward measurements and V_r (reverse voltage at current density J = 0.5 mA/cm^2).

Frame Name, Diode Size (µm)	n	ϕ_B (Volt)	V_r (V)
Frame 1, 200	1.08	0.74	−110 V
Frame 1, 100	1.03	0.80	−125 V
Frame 1, 50 (high)	1.07	0.83	−120 V
Frame 1, 50 µm (low)	1.11	0.83	−150 V
Frame 2, 200	1.17	0.72	−70 V
Frame 2, 100	1.09	0.78	−125 V
Frame 2, 50 (high)	1.13	0.82	NONE
Frame 2, 50 (low)	1.09	0.82	−155 V
Frame 3, 200	1.03	0.83	−180 V
Frame 3, 100	1.01	0.87	−170 V
Frame 3, 50 (high)	1.08	0.88	−175 V
Frame 3, 50 µm (low)	1.07	0.89	−180 V
Frame 4, 200	1.03	0.84	−185 V
Frame 4, 100	1.09	0.85	−160 V
Frame 4, 50 µm (high)	1.00	0.91	−180 V
Frame 4, 50 µm (low)	1.00	0.91	−175 V

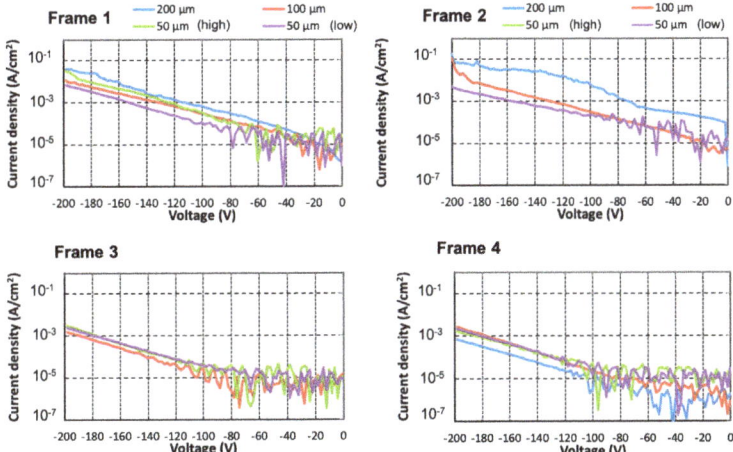

Figure 8. Reverse I-V characteristics for diodes from the frames (1, 2, 3 and 4): blue (200 µm diode), red (100 µm diode), green (50 µm high), purple (50 µm diode low).

4. Discussion

The E_2^h Raman position maps correlate well with the cathodo-luminescence measurements (see Figure 4), showing that the Raman spectroscopy can be used to localize the threading dislocations in the GaN active layer. Our measurements reveal that the dislocation clusters may locally decrease the stress in a similar way as the dislocations associated with the grain boundaries. In addition, the 2D $A_1(LO)$ position and intensities maps were used to estimate the carrier concentration in the diodes and identify peculiar defects. Figures 5 and 6 show that the area occupied by the defects is particularly important with respect to the diode size for the two 200 μm diodes of frames 1 and 2. These peculiar defects locally decrease the net carrier concentration and therefore may possibly affect the Schottky diode electrical performances. The SIMS results highlighted the presence of a significant amount of oxygen and carbon, and the changes in the incorporation of silicon or of these impurities could explain the observed variation of the *n*-carrier concentration. The I-V forward characterizations show that all the studied diodes have a typical Schottky behavior with ideality factors between 1 and 1.17. The I-V reverse characterizations show that all the studied diodes have leakage currents increasing with the applied reverse voltage and they are higher for the diodes of frame 1 and 2, especially for the two 200 μm diodes. These results seem to indicate that the defects observed in the A_1(LO) Raman maps (Figures 5 and 6) are electrically active and contribute to the reverse leakage. Thus, the electrical leakage in the reverse biased diodes seems more correlated with the short range non-uniformities of the effective doping rather than with the strain fluctuation induced by the dislocations, a result in agreement with our previous investigations on dislocation clusters [12].

5. Conclusions

In this work, vertical GaN Schottky diodes on a freestanding GaN substrate were fabricated and investigated. The physical and the electrical properties were studied with cathodo-luminescence, micro-Raman mapping, SIMS measurements, and current-voltage (I-V) to understand the effects of physical parameters such as threading dislocations and doping concentration homogeneity on the electrical performances of SBDs. Evidence of dislocations in the diode epilayer was spotted thanks to CL measurements and a correlation with 2D mappings of the Raman E_2^h signal was observed. The I-V measurements of the diodes reveal a significant increase in the leakage current with applied reverse bias up to 200 V. However, no clear correlation with the presence of dislocation cluster in the area occupied by the diodes and excessive leakage current were observed. On the contrary, the mapping of the A1 (LO) Raman mode that is sensitive to fluctuations of the effective doping level shows correlations with the leakage. Even when the origin of such fluctuations remains to be identified, this study shows the efficiency of micro-Raman spectroscopy to probe structural and electronic properties of GaN-based electrical vertical devices. The combination of physical and electrical characterization methods indicates that the electrical leakage in the reverse biased diodes seems more correlated with short range non-uniformities of the effective doping than with the strain fluctuation induced by the dislocations and thus, this method can be efficient to study vertical GaN electrical power devices. Finally, we suggest further investigations such as DLTS measurements to set up a solid correlation between physical properties and the observed electrical behavior of the diodes [13].

Author Contributions: Data curation, A.J.E.N., C.S. and S.S.; formal analysis, A.J.E.N., S.S. and T.H.N.; investigation, A.J.E.N., S.S., T.H.N., P.D.M., E.F. and Y.C.; methodology, C.S., Y.C. and D.P.; project administration, L.V.P., F.M., Yvon Cordier and H.M.; supervision, C.S., H.M. and D.P.; validation, C.S. and D.P.; writing—original draft, A.J.E.N.; writing—review and editing, A.J.E.N., C.S., Y.C., L.V.P., F.M., H.M. and D.P. All authors have read and agreed to the published version of the manuscript.

Funding: This work was supported by the French Technology Facility Network RENATECH, the French National Research Agency (ANR) through the projects C-Pi-GaN (Grant No. ANR-18- CE05-0045), GaNeX (ANR-11-LABX-0014) and the AuRA Region (Région Auvergne-Rhône Alpes) through the OptiGaN project.

Data Availability Statement: The data presented in this study are available on request from the corresponding author.

Conflicts of Interest: The authors declare no conflict of interest.

References

1. Cao, Y.; Chu, R.; Li, R.; Chen, M.; Chang, R.; Hughes, B. High-voltage vertical GaN Schottky diode enabled by low-carbon metal-organic chemical vapor deposition growth. *Appl. Phys. Lett.* **2016**, *108*, 62103. [CrossRef]
2. Yeluri, R.; Lu, J.; Hurni, C.A.; Browne, D.A.; Chowdhury, S.; Keller, S.; Speck, J.S.; Mishra, U.K. Design, fabrication, and performance analysis of GaN vertical electron transistors with a buried p/n junction. *Appl. Phys. Lett.* **2015**, *106*, 183502. [CrossRef]
3. Amano, H.; Baines, Y.; Beam, E.; Borga, E.; Bouchet, T.; Chalker, P.R.; Charles, M.; Chen, K.J.; Chowdury, N.; Chu, R.; et al. The 2018 GaN power electronics roadmap. *J. Phys. D Appl. Phys.* **2018**, *51*, 163001. [CrossRef]
4. Ren, B.; Liao, M.; Sumiya, M.; Wang, L.; Koide, Y.; Sang, L. Nearly ideal vertical GaN Schottky barrier diodes with ultralow turn-on voltage and on-resistance. *Appl. Phys. Express* **2017**, *10*, 51001. [CrossRef]
5. Sang, L.; Ren, B.; Sumiya, M.; Liao, M.; Koide, Y.; Tanaka, A.; Cho, Y.; Harada, Y.; Nabatame, T.; Sekiguchi, T.; et al. Initial leakage current paths in the vertical-type GaN-on-GaN Schottky barrier diodes. *Appl. Phys. Lett.* **2017**, *111*, 122102. [CrossRef]
6. Tompkins, R.P.; Walsh, T.A.; Derenge, M.A.; Kirchner, K.W.; Zhou, S.; Nguyen, C.B.; Jones, K.A. The effect of carbon impurities on lightly doped MOCVD GaN Schottky diodes. *J. Mater. Res.* **2011**, *26*, 2895–2900. [CrossRef]
7. Fujikura, H.; Hayashi, K.; Horikiri, F.; Narita, Y.; Konno, T.; Yoshida, T.; Ohta, H.; Mishima, T. Elimination of macrostep-induced current flow nonuniformity in vertical GaN PN diode using carbon-free drift layer grown by hydride vapor phase epitaxy. *Appl. Phys. Lett. Express* **2018**, *11*, 45502. [CrossRef]
8. Usami, S.; Ando, Y.; Tanaka, A.; Nagamatsu, K.; Deki, M.; Kushimoto, M.; Nitta, S.; Honda, Y.; Amano, H.; Sugawara, Y.; et al. Correlation between dislocations and leakage current of p-n diodes on a free-standing GaN substrate. *Appl. Phys. Lett.* **2018**, *112*, 182106. [CrossRef]
9. Amilusik, M.; Wlodarczyk, D.; Suchocki, A.; Bockowski, M. Micro-Raman studies of strain in bulk GaN crystals grown by hydride vapor phase epitaxy on ammonothermal GaN seeds. *Jpn. J. Appl. Phys.* **2019**, *58*, SCCB32. [CrossRef]
10. Kozawa, T.; Kachi, T.; Kano, H.; Taga, Y.; Hashimoto, M.; Koide, N.; Manabe, K. Raman scattering from LO phonon-plasmon coupled modes in gallium nitride. *J. Appl. Phys.* **1994**, *75*, 1098. [CrossRef]
11. Gogova, D.; Larsson, H.; Kasic, A.; Yazdi, G.R.; Ivanov, I.; Yakimova, R.; Monemar, B.; Aujol, E.; Frayssinet, E.; Faurie, J.-P.; et al. High-Quality 2″ Bulk-Like Free-Standing GaN Grown by HydrideVapour Phase Epitaxy on a Si-doped Metal Organic Vapour Phase Epitaxial GaN Template with an Ultra Low Dislocation Density. *Jpn. J. Appl. Phys.* **2005**, *44*, 3R. [CrossRef]
12. Ngo, T.H.; Comyn, R.; Frayssinet, E.; Chauveau, H.; Chenot, S.; Damilano, B.; Tendille, F.; Cordier, Y. Cathodoluminescence and electrical study of vertical GaN-on-GaN Schottky diodes with dislocation clusters. *J. Cryst. Growth* **2020**, *552*, 125911.
13. Vigneshwara Rajaa, P.; Raynaud, C.; Sonneville, C.; N'Dohi, A.E.; Morel, H.; Phung, L.V.; Ngo, T.H.; De Mierry, P.; Frayssinet, E.; Maher, H.; et al. Comprehensive characterization of vertical GaN-on-GaN Schottky barrier diode. *Microelectron. J.* **2022**, *128*, 105575. [CrossRef]
14. Reshchikov, M.A.; Vorobiov, M.; Andrieiev, O.; Ding, K.; Izyumskaya, N.; Avrutin, V.; Usikov, A.; Helava, H.; Makarov, Y. Determination of the concentration of impurities in GaN from photoluminescence and secondary-ion mass spectrometry. *Sci. Rep.* **2020**, *10*, 2223. [CrossRef]
15. Freitas, J.A., Jr.; Moore, W.J.; Shanabrook, B.V.; Braga, G.C.; Lee, S.K.; Park, S.S.; Han, J.Y.; Koleske, D.D. Donors in hydride-vapor-phase epitaxial GaN. *J. Cryst. Growth* **2002**, *246*, 307–314. [CrossRef]
16. Popovici, G.; Kim, W.; Botchkarev, A.; Tang, H.; Morkoc, H. Impurity contamination of GaN epitaxial films from the sapphire, SiC and ZnO substrates. *Appl. Phys. Lett.* **1997**, *71*, 3385–3387. [CrossRef]
17. Ciarkowski, T.; Allen, N.; Carlson, E.; McCarthy, R.; Youtsey, C.; Wang, J.; Fay, P.; Xie, J.; Guido, L. Connection between Carbon Incorporation and Growth Rate for GaN Epitaxial Layers Prepared by OMVPE. *Materials* **2019**, *12*, 2455. [CrossRef]
18. Wagner, J.-M.; Bechstedt, F. Phonon deformation potentials of α-GaN and -AlN: An ab initio calculation. *Appl. Phys. Lett.* **2000**, *77*, 346–348. [CrossRef]
19. Demangeot, F.; Frandon, J.; Baules, P.; Natali, F.; Semond, F.; Massies, J. Phonon deformation potentials in hexagonal GaN. *Phys. Rev. B* **2004**, *69*, 155215. [CrossRef]
20. Wagner, J.-M.; Bechstedt, F. Properties of strained wurtzite GaN and AlN: Ab initio studies. *Phys. Rev. B* **2002**, *66*, 115202. [CrossRef]
21. Kokubo, N.; Tsunooka, Y.; Fujie, F.; Ohara, J.; Onda, S.; Yamada, H.; Shimizu, M.; Harada, S.; Tagawa, M.; Ujihara, T. Non-destructive visualization of threading dislocations in GaN by micro raman mapping. *Jpn. J. Appl. Phys.* **2019**, *58*, SCCB06. [CrossRef]
22. Belabbas, I.; Béré, A.; Chen, J.; Ruterana, P.; Nouet, G. Investigation of the atomic core structure of the (a and c)-mixed dislocation in wurtzite GaN. *Phys. Stat. Solid C* **2007**, *4*, 2940–2944. [CrossRef]
23. Dadgar, A.; Poschenrieder, M.; Reiher, A.; Bläsing, J.; Christen, J.; Krtschil, A.; Finger, T.; Hempel, T.; Diez, A.; Krost, A. Reduction of stress at the initial stages of GaN growth on Si(111). *Appl. Phys. Lett.* **2003**, *82*, 28. [CrossRef]
24. Nenstiel, C.; Bügler, M.; Callsen, G.; Nippert, F.; Kure, T.; Fritze, S.; Dadgar, A. Germanium–The superior dopant in n-type GaN. *Phys. Stat. Solidi (RRL)–Rapid Res. Lett.* **2015**, *9*, 716–721. [CrossRef]
25. Artús, L.; Cusco, R.; Ibanez, J.; Blanco, N.; Gonzalez-Diaz, G. Raman scattering by LO phonon-plasmon coupled modes in n-type InP. *Phys. Rev. B* **1999**, *60*, 5456–5463. [CrossRef]
26. Kuball, M. Raman spectroscopy of GaN, AlGaN and AlN for process and growth monitoring/control. *Surf. Interface Anal.* **2001**, *31*, 987–999. [CrossRef]

27. Peng, Y.; Xu, X.; Hu, X.; Jiang, K.; Song, S.; Gao, Y.; Xu, H. Raman spectroscopic study of the electrical properties of 6H–SiC crystals grown by hydrogen-assisted physical vapor transport method. *J. Appl. Phys.* **2010**, *107*, 93519. [CrossRef]
28. N'Dohi, A.E.; Sonneville, C.; Phung, L.V.; Ngo, T.H.; De Mierry, P.; Frayssinet, E.; Maher, H.; Tasselli, J.; Isoird, K.; Morancho, F.; et al. Micro-Raman characterization of homo-epitaxial n doped GaN layers for vertical device applications. *AIP Adv.* **2022**, *12*, 25126. [CrossRef]
29. Bandić, Z.Z.; Bridger, P.M.; Piquette, E.C.; McGill, T.C. The values of minority carrier diffusion lengths and lifetimes in GaN and their implications for bipolar devices. *Solid-State Electron.* **2000**, *44*, 221–228. [CrossRef]
30. Pugatschow, A.; Heiderhoff, R.; Balk, L.J. Time resolved determination of electrical field distributions within dynamically biased power devices by spectral EBIC investigations. *Microelectron. Reliab.* **2007**, *47*, 1529–1533. [CrossRef]
31. Sawada, M.; Sawada, T.; Yamagata, Y.; Imai, K.; Kimura, H.; Yoshino, M.; Iizuka, K.; Tomozawa, H. Electrical characterization of n-GaN Schottky and PCVD-SiO2/n-GaN interfaces. *J. Cryst. Growth* **1998**, *189–190*, 706–710. [CrossRef]

Disclaimer/Publisher's Note: The statements, opinions and data contained in all publications are solely those of the individual author(s) and contributor(s) and not of MDPI and/or the editor(s). MDPI and/or the editor(s) disclaim responsibility for any injury to people or property resulting from any ideas, methods, instructions or products referred to in the content.

Article

Normally-Off p-GaN Gate High-Electron-Mobility Transistors with the Air-Bridge Source-Connection Fabricated Using the Direct Laser Writing Grayscale Photolithography Technology

Yujian Zhang [1,2], Guojian Ding [2], Fangzhou Wang [2,3], Ping Yu [2], Qi Feng [2], Cheng Yu [1,2], Junxian He [1,2], Xiaohui Wang [2], Wenjun Xu [2], Miao He [1,*], Yang Wang [2,*], Wanjun Chen [3], Haiqiang Jia [2,4] and Hong Chen [2,4]

[1] School of Physics and Optoelectronic Engineering, Guangdong University of Technology, Guangzhou 510006, China; 2112015107@mail2.gdut.edu.cn (Y.Z.)
[2] Songshan Lake Materials Laboratory, Dongguan 523808, China
[3] State Key Laboratory of Electronic Thin Films and Integrated Devices, University of Electronic Science and Technology of China, Chengdu 610054, China
[4] Key Laboratory for Renewable Energy, Institute of Physics, Chinese Academy of Sciences, Beijing 100190, China
* Correspondence: herofate@gdut.edu.cn (M.H.); wangyang@sslab.org.cn (Y.W.)

Abstract: In this work, we used the Direct Laser Writing Grayscale Photolithography technology to fabricate a normally-off p-GaN gate high-electron-mobility transistor with the air-bridge source-connection. The air-bridge source-connection was formed using the Direct Laser Writing Grayscale Photolithography, and it directly connected the two adjacent sources and spanned the gate and drain of the multi-finger p-GaN gate device, which featured the advantages of stable self-support and large-span capabilities. Verified by the experiments, the fabricated air-bridge p-GaN gate devices utilizing the Direct Laser Writing Grayscale Photolithography presented an on-resistance of 36 $\Omega \cdot$mm, a threshold voltage of 1.8 V, a maximum drain current of 240 mA/mm, and a breakdown voltage of 715 V. The results provide beneficial design guidance for realizing large gate-width p-GaN gate high-electron-mobility transistor devices.

Keywords: p-GaN gate HEMT; air-bridge source-connection; Direct Laser Writing Grayscale Photolithography technology

Citation: Zhang, Y.; Ding, G.; Wang, F.; Yu, P.; Feng, Q.; Yu, C.; He, J.; Wang, X.; Xu, W.; He, M.; et al. Normally-Off p-GaN Gate High-Electron-Mobility Transistors with the Air-Bridge Source-Connection Fabricated Using the Direct Laser Writing Grayscale Photolithography Technology. Crystals 2023, 13, 815. https://doi.org/10.3390/cryst13050815

Academic Editor: Evgeniy N. Mokhov

Received: 13 April 2023
Revised: 8 May 2023
Accepted: 10 May 2023
Published: 13 May 2023

Copyright: © 2023 by the authors. Licensee MDPI, Basel, Switzerland. This article is an open access article distributed under the terms and conditions of the Creative Commons Attribution (CC BY) license (https://creativecommons.org/licenses/by/4.0/).

1. Introduction

Gallium nitride- (GaN) based high-electron-mobility transistors (HEMT) are promising candidates for high-power applications due to their superior material and device characteristics [1,2]. Two-dimensional electron gas (2DEG) is created and limited at the interface between AlGaN/GaN by the piezoelectric effect of the GaN/AlGaN heterojunction and the spontaneous polarization effect. Based on the conventional structure of AlGaN/GaN-channeled electrons, it allows only the construction of depletion-mode normally-on devices. D-mode devices must import a continuous negative gate voltage to realize device turn-off; hence, they need extra-complicated gate-driver circuitry. In order to achieve normally-off GaN HEMT, different technologies have been proposed, such as cascode configuration combining Si-based normally-off MOSFET and GaN-based normally-on HEMT, fluorine-ion-implantation gate HEMT, gate-recessed MIS (metal–insulator–semiconductor)-HEMT, and p-GaN gated HEMT. Among the various GaN devices, p-GaN gate HEMT have the advantage of a stable threshold voltage and high repeatability of the process and have been commercially accepted and adopted [3]. With the development of the high-current p-GaN gate HEMT, the source-connection technology has been utilized to realize the large gate-width device [4]. Several common interconnection methods are adopted in multi-finger devices: the dielectric-isolation connection, back-hole connection using Through Silicon Via (TSV), and air-bridge connection. The air-bridge source-connection is an effective method to connect the source in the large gate-width device with a multi-finger structure [5]. The source-connected air-bridge field plate (AFP), which jumps from the source over the gate and lands between the gate and drain, has been fabricated. Compared to HEMT with a conventional field plate, this AFP structure can result in a three-times-higher improvement in forward-blocking voltage, attain 375 V, 37% lower $R_{on,sp}$, and drain leakage current lower by one order of magnitude. Dora Y shows that slant-field-plate technology has been fabricated and proven to be very effective; the slant-field plate is self-aligned with the gate in a single-process step and is shown to support a breakdown voltage up to 1900 V [6]. The Coffie R model shows that transition angles of less than 30° (measured from the surface) can result in significant improvements in electric field management [7].

In this work, we demonstrated an air-bridge source-connection p-GaN gate HEMT, fabricated using Direct Laser Writing Grayscale Photolithography (DLWGP) technology. Through controlling a different intensity or dose of light, the DLWGP can expose different vertical depths of the photoresist of the same thickness to quickly obtain the air-bridge source-connection structure. Compared to the E-beam [8,9] and digital micromirror device (DMD) [10] that are reported so far, the DLWGP draws more attention because it is maskless, versatile for the various semiconductor materials, suitable for the fast iteration of prototypes, and so on. By utilizing the DLWGP technology, the multi-finger p-GaN gate HEMT with the air-bridge source-connection is well-presented. The fabrication processes are discussed, and the performances of the experimental device are characterized and analyzed.

2. Experiments

Two methods have been reported to realize the p-GaN gate structure. One is the selective epitaxy p-GaN layer on the surface of AlGaN barrier [11]. Interface pollution and hole concentration are major problems for selective epitaxy and result in this technology being difficult to widely fabricate at present. The other one is whole epitaxy, growing a whole p-GaN layer on the surface of AlGaN and removing p-GaN layers except for the gate, using etching with which a channel of 2DEG could reappear except for the gate. Adopting this method to realize p-GaN gate structure has the advantage of stable threshold voltage and a repeatable process of fabrication, and it is widely applied to fabricate E-mode HEMT.

In this paper, we fabricated an enhanced HEMT device by adopting the etching p-GaN method. The process modules for the (E-mode) GaN HEMT device fabrication included device isolation, p-GaN gate formation, source and drain ohmic contact, gate Schottky contact, surface passivation, and DWLGP air-bridge connection (as shown in Figure 1). We will focus on several key process technologies for p-GaN HEMT.

The p-GaN/AlGaN/GaN heterostructures were grown using MOCVD on 800 μm Si substrate. The layer stack consisted of a 2 nm AlN nucleation, a 4 μm (Al)GaN buffer, a 200 nm GaN channel, a 15 nm AlGaN barrier (with Al content of 0.2), and a 70 nm thickness Mg-doped p-GaN cap layer with a doping concentration of 1×10^{18} cm^{-3}.

2.1. Device Isolation

Device isolation is a step in the fabrication process to separate adjacent devices by destroying the electronic channel and reducing the impact of current leakage between devices. There are two common methods to realize device isolation. The first method is by adopting ion-implantation equipment to isolate the active area of the p-GaN HEMT device (implantation ion species including B+, N+, and He+). In this way, the channel of the heterojunction interface is destroyed with incident ions and realizes the isolation of devices. The ion-implantation process generally adopts several steps (implant energy gradually increasing). The advantage of the ion-implantation method is that the device surface stays flat after ion implantation, and the process is simple. The disadvantages are the high expense of the equipment, and that the injection energy is too large, which will cause the denaturation of the photoresist, and the subsequent removal of the denaturation of the photoresist raises the high requirements.

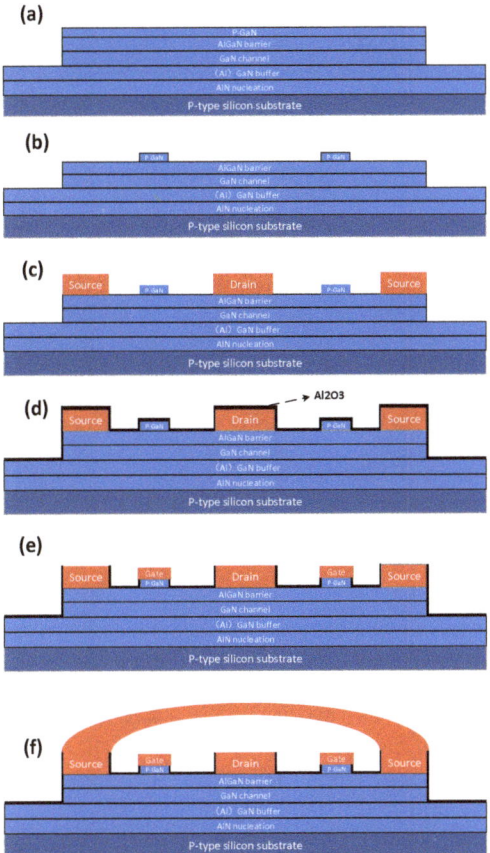

Figure 1. Fabrication processes of the air-bridge common-source normally-off Mg-GaN/AlGaN/GaN HEMT (**a**) mesa etching; (**b**) p-GaN etching; (**c**) evaporated Ti/Al/Ni/Au and annealing treatment (**d**) ALD-Al$_2$O$_3$ passivation/dielectric; (**e**) removal of the top of the p-GaN, drain and source passivation, and evaporated Ni/Au and pad metal. (**f**) air-bridge common-source using grayscale technology.

The second method adopts Inductively Coupled Plasma dry etching to destroy GaN channels outside the active area (requiring an etching depth far greater than the channel depth). After that, there is the forming of a mesa structure with a large height difference and without an electronic channel around it to achieve separation and isolation between devices. Among them, the Reactive Ion Etching (RIE) and Inductively Coupled Plasma (ICP) dry etching machines are the most widely used etching methods in GaN materials. The advantages of adopting a dry etching method to form mesa isolation are the simple process, the fact that it is easy to realize, and the low expense of the equipment. GaN dry etching using ICP has a physical bombardment and chemical reaction in which a chlorine base plays a major role in the chemical reaction.

In addition to the choice of chemistries employed, etching features and etching rate are affected by the gas flow, antenna RF Power, bias RF power, chamber pressure, and so forth. The chamber pressure influences the etching rate. Increasing the chamber pressure (higher than 30 mTorr) will cause the mean free path of the reactive molecules to shorten, plasma kinetic energy to reduce, and the physical bombardment etching rate to decline. Adopting the dry etching mesa method achieves isolation, and the adjacent device relies on the big resistivity of the buffer-layer material. When using the dry etching mesa for isolation, it should be noted that the thickness of the photoresist at the edge of the mesa is less than the flat photoresist, which will cause difficulty in metal lift-off. In the process of mesa etching, the corresponding marks are prepared together with the mesa etching. The area between the marks and the surrounding etched surface forms an obvious height difference to provide mark recognition for later alignment. This process is unsuitable for metal to mark because subsequent high-temperature annealing above 800 °C is required. The high temperature will damage the shape of the mark, which is inconvenient for the subsequent alignment.

2.2. p-GaN Gate Formation

Due to the need to achieve enhanced devices (normally-off HEMT), there are four ways to achieve enhanced devices described. P-GaN gate achieves enhanced HEMT; this device was adopted due to stable threshold voltages, which p-GaN gate had shown great potential for in GaN power device applications, and it had been commercialized [12]. The p-GaN layer lifts the energy bands of AlGaN, causing the 2DEG to disappear. In the process of fabricating the p-GaN gate device, the p-GaN layer other than the gate region was removed using ICP dry etching, and 2DEG reappeared. P-GaN etching is the key process in the fabrication of p-GaN gate devices and precisely controls the etching depth of the p-GaN cap layer as necessary. The residual p-GaN layer will deplete the 2DEG density, resulting in a decrease in the current density. Likewise, the over-etching of the AlGaN barrier layer will also decrease the current density due to it decreasing the polarization effect [13]. Therefore, the mesa etching recipe was unsuitably used for p-GaN etching. We need to optimize the p-GaN/AlGaN etching selection ratio. Two methods are recommended to achieve selective dry etching. One is fluorine-based chemical etching and the other is oxygen-based chemical etching. Both achieve a large p-GaN/AlGaN etching ratio by reducing the AlGaN etching rate.

Chang YC showed that adopting the $Cl_2/BCl_3/SF_6$ mixed gas plasma self-terminating dry etching technique fabricated normally-off p-GaN HEMT devices [14]. This etching technique features accurate etching depth control and low surface-plasma damage. When SF_6 reaches the AlGaN layer, fluorine-based gas reacts with Al and forms non-volatility $AlF3$ on the AlGaN surface, which significantly reduces the etching rate of the AlGaN layer, thus achieving selective etching of p-GaN over AlGaN. After self-terminating etching, the thin AlF3 layer on the surface was removed using a Buffered Oxide Etchant (BOE).

The other option is adopting oxygen and chlorine-based mixed gas. Taube A presented the results of the development of the selective etching of p-GaN over AlGaN in $Cl_2/Ar/O_2$ ICP plasma for the fabrication of normally-off p-GaN gate GaN HEMT using a laser reflectometry system for the precise control of the etched-material thickness [15]. By optimizing etching process parameters such as oxygen flow, ICP power, and chamber pressure, the high-etching selectivity of p-GaN and AlGaN was obtained, with values up to 56:1. The formation of thin Al_2O_3 surface passivation layers on the AlGaN surface was in an atmosphere containing oxygen, which is strongly resistant to etching and leads to a low etching rate of AlGaN. Increasing O_2 flow (0–5 sccm) resulted in a decrease in etching rates for both GaN and AlGaN. The selectivity was increased with an O_2 flow range of 0–2 sccm, but an O_2 flow of more than 3 sccm while etching caused selectivity to begin to descend.

In this paper, we used a $Cl_2/O_2/Ar$ mixture gas for the selective etching of the p-GaN layer. Firstly, we used Plasma Enhanced Chemical Vapor Deposition (PECVD) to deposit a 200 nm SiO_2 protective film in the active region. The SiO_2 film above the gate region was protected by the photoresist, and the non-gate region SiO_2 was etched using CF_4-based ICP. After removing the photoresist, the p-GaN was removed except in the gate region using the high-selection ratio of mixed-gas-based ICP. Finally, we soaked the BOE to remove the SiO_2 protective film above the gate.

2.3. Ohmic Contacts

The source and drain of the Ohmic contact are the key steps in the preparation of GaN HEMT. A typical Ohmic contact is widely adopted using an electron beam to vaporize four layers of Ti/Al/Ni/Au metals and quickly anneal under the protection of nitrogen or other inert gases at high temperatures above 800 °C. Among them, the Ti-based layer metal plays a dominant role and reacts with the GaN surface N to form TiN at high temperatures, increasing the nitrogen vacancy at the interface, which leads to the thinning of the electron tunneling barrier [16]. However, single-layer Ti is easy to oxidize, so it is necessary to add Au metal to protect Al metal from oxidizing. Due to the strong downward diffusion of Au metal, a layer of Ni metal is inserted as a metal barrier layer in order to prevent the Au from diffusing with the Ti/Al. The Ohmic contact formed using this method is of low resistance and is easy to implement. Adopting Au-free metal stacking to make Ohmic contacts is strongly pursued at present as Au-free reduces expense, and Au is incompatible with the CMOS process. W Liang et al. presented a low Ohmic-contacts resistance formed with sidewall contacts with an Ohmic recess. Due to the direct sideways contact, which reduced the resistance of tunneling through the AlGaN barrier, it was believed to be a more efficient carrier transport mechanism [17].

In this paper, we adopted a lift-off metal method to fabricate source and drain ohmic metal. Coating the photoresist, the photoresist of the source and drain was removed using exposure and developer. The source and drain of the AlGaN barrier layer were thinned using Cl_2-based ICP before metal evaporation to reduce contact resistance. The metal stack of Ti/Al/Ni/Au was evaporated in the source and drain region using electron beam evaporation. Soaking in acetone lifted off non-source/drain metal. Finally, annealing at 850 °C for 30 s in the ambient of N_2 was performed to form the Ohmic contact.

2.4. Surface Passivation

The passivation layer is an essential step in the HEMT power-device process to protect the device from external influences and repair device performance. GaN device surface state is a complicated problem. These surface states may come from suspended bonds of surface atoms, plasma damage during etching p-GaN, surface pollutants, and so on. Vetury R showed surface states in the vicinity of the gate trap electrons and the depletion of channel electrons, which thus acted as a negatively charged virtual gate, which would cause current collapse [18]. A suitable passivation layer is needed to stabilize device performance and reliability. Plasma Enhanced Chemical Vapor Deposition (PECVD), Atomic Layer Deposition (ALD), and Low Pressure Chemical Vapor Deposition (LPCVD) are commonly used to prepare passivation layers of devices. At present, it has been explored that the insulation material widely used in the passivation layer of HEMT devices, including SiO_2, Si_3N_4 [19], and Al_2O_3 [20]. In addition, GaN can be oxidized in the air forming a layer of gallium sub-oxide (GaO_x) on the surface. Its oxidized defect surface state will lead to a gate leakage current. Bae. C proposed a plasma-assisted oxidation method before PECVD to significantly reduce defect state densities [21]. Compared with directly growing the passivation layer, plasma-assisted oxidation before the PECVD method reduced the gate-leakage current because adequate plasma oxidation transformed gallium sub-oxide (GaO_x) into high-quality insulation materials.

In this paper, we deposited a high-quality Al_2O_3 passivation layer through ALD. Due to the ALD Self-limiting Reaction characteristic, the film of the Al_2O_3 passivation layer was deposited one by one. The main advantage of the ALD compared to other equipment is the high coverage and quality to better isolate the device from the external environment. We could control the thickness of the passivation layer by controlling the number of layers deposited.

2.5. Schottky Gate Metal

With normally-off devices with p-GaN layers, we usually steamed high-work function metal on the gate to form Schottky contacts. The difference in metal and semiconductor barrier height depends on the difference between the work function of the metal and the electron affinity of the semiconductor. With fixed gate voltage, the leakage current of the gate can be reduced effectively by increasing the height of the Schottky barrier. Therefore, some high-work function metals such as Ni, Ti, and W were used to prepare the Schottky contact-based metal. Single Ni metal is easily oxidized by air, so it is necessary to steam Au metal behind Ni to protect against oxidation while reducing the gate resistance.

Lu X et al. reported adopting a two-step process combining a pre-gate surface treatment and post-gate annealing to reduce the off-state leakage current in HEMT [22]. The post-gate annealing process reduced the vertical tunneling leakage current by improving the Schottky contact quality of the transistor gate. The device's off-state leakage current was reduced by about 7 orders.

Considering that the Schottky metal needs better adhesion on the p-GaN surface to prevent the metal from falling off in the subsequent process, we chose the stack of Ni/Au metal as the Schottky metal. After we deposited the passivation layer, we coated the photoresist on the surface of the passivation layer and removed the upper photoresist of source/drain and p-GaN using exposing and developing. Then, we used Cl_2-based ICP to remove the upper of p-GaN Al_2O_3 passivation. The stack of Ni/Au gate metals was deposited above the p-GaN gate to form the Schottky contact using Electron beam evaporation while the photoresist protected the non-gate regions.

2.6. Pad Metal

In order to conveniently test the performance of the ten-finger gate device, we needed to connect all the gates and drains of the device to a large enough Pad metal through metal lines. It is worth noting that the surface of Ohmic metal was too rough after annealing, and we cautiously used Ohmic metal as the Pad metal, which can cause bad contact between the probe and Pad during testing. Therefore, we prepared new Pad metal of source, drain, and gate using Electron Beam Evaporation while the drain and gate of the device were, respectively, connected to the corresponding Pad.

2.7. Air-Bridge Source-Connection with Direct Laser Writing Grayscale Photolithography

In the preceding process, the gate and drain electrode of devices had been connected to the corresponding Pad metal with metal lines. In this case, the source electrode of the device was without a connected path to the source Pad metal, so it needed to be connected from the second plane to avoid a short circuit. In this paper, the source of the device was directly connected to the pad from the top of the device using air-bridge-connect technology. In this paper, we proposed that fabricated air-bridges were adopted with the grayscale exposure method. Compared with other methods, the grayscale exposure method can fabricate a more flexible vertical design of air-bridges, and the fabrication of the grayscale exposure method is simpler.

In this work, we adopted the Heidelberg DWL66+ equipment grayscale-exposure mode to fabricate the air-bridge source-connection. This DWL66+ supports a minimum structure size of 0.3 μm, 0.6 μm, and 1 μm. The 0.3 μm minimum-size writing speed corresponded to 3 mm^2/min, and the 0.6 μm minimum-size writing speed corresponded to 13 mm^2/min. With both more quickly obtaining a 3D structure and supplying the demand of the design minimum line, we adopted a 0.6 μm minimum line-width gray mode. Standard and advanced grayscale modes were 128/256 gray levels, respectively. Moreover, the DWL66+ output a 405 nm light source using a diode laser (300 mw maximum output power). This DWL66+ supplied the air-gauge or optical autofocus for the exposure of small samples (less than 10 mm).

According to the DLWGP exposure principle, the upper part of the photoresist is first illuminated; hence, some 3D structures are impossible to achieve using the grayscale-exposure mode. The photoresist structure of the air-bridge support layer met the requirements for grayscale lithography, which can be decomposed into three parts, as shown in Figure 2a: pier, bridge photoresist, and the surrounding photoresist areas. Therein, the pier part must be fully exposed, and the part of the span bridge with curved exposure energy and the surrounding sacrificial photoresist part are unexposed in DLWGP mode. When designing the gray-exposure map of the air-bridge structure, we can simply divide the 3D structure into a 2D distribution map and height value. At present, the mainstream layout software is not compatible with gray value design, so we needed to design the bmp map through another pixel software. When we design a bmp map, we must follow the DLWGP conversion graph paper rule (one pixel is equal to 100 nm, and the center of the pixel map corresponds to the center of exposure) to plan the air-bridge structure distribution map; otherwise, we are unable to align the previous layer. In addition to locating the location of the air-bridge, we also need to set the gray value of the bridge and pier. In DWLGL mode, this device will divide the laser energy into 256 equivalencies (corresponding to 256 gray steps); the 255 value corresponds to the full value of the exposure power, and the 0 value corresponds to no exposure. Based on the criterion that the photoresist thickness of 6.9 μm corresponds to every grayscale value of 27 nm, the bridge camber is realized by designing a curve gradient gray value. After designing the bmp diagram, we can adjust the DWLGP.

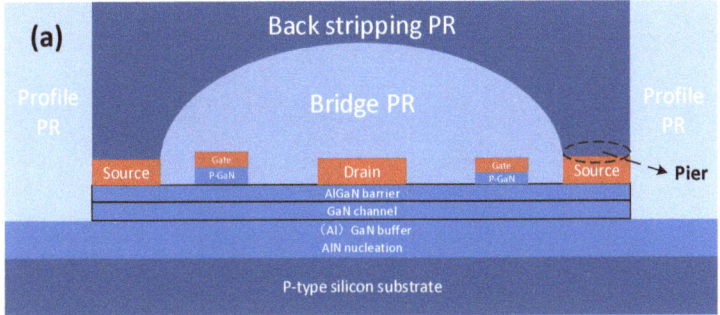

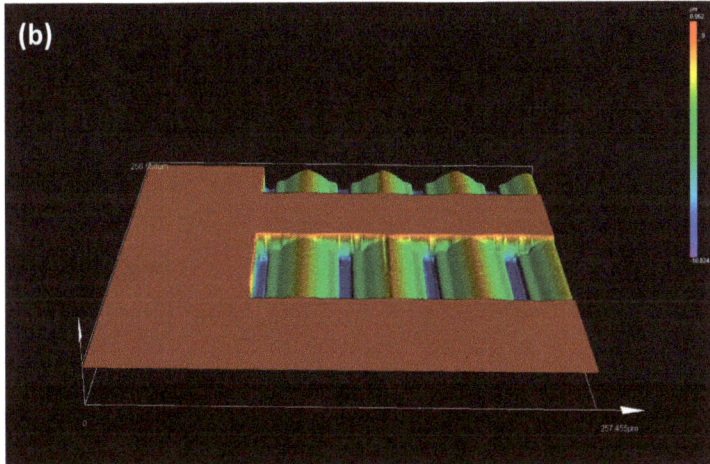

Figure 2. (a) grayscale exposure photoresist chart (b) laser scanning confocal microscope view 3D photoresist structure.

In order to obtain a metal bridge of sufficient thickness, we spun a coated target thickness of 6.9 μm AZ4562 photoresist. The appropriate gray exposure parameters were used to mold the photoresist for 3D exposure. After exposure, the photoresist was immersed in the newly configured developer (1:4/400 K: water) for 4 min. The developer removed the upper layer of the sensitive photoresist, and the 3D photoresist morphology could be scanned using a laser confocal microscope, as shown in Figure 2b. We could measure the line-width at the pier to meet our design requirement that the grid is still protected by the photoresist. Then we evaporated Ti/Al 20/1500 nm metals through electron beam deposition. It was soaked in acetone for 1 h, and the surrounding metals were taken off as the photoresist dissolved in the acetone. The remaining metal bridge structure (Figure 3) was observed through SEM, of which the air-bridge spans the grid and drain above to connect adjacent sources.

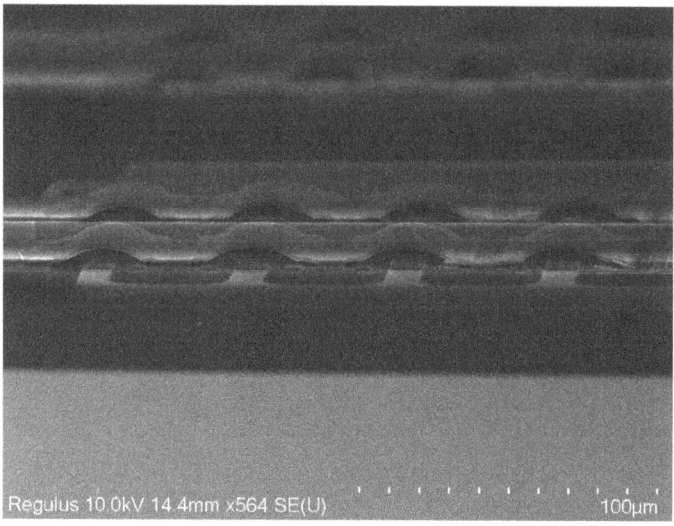

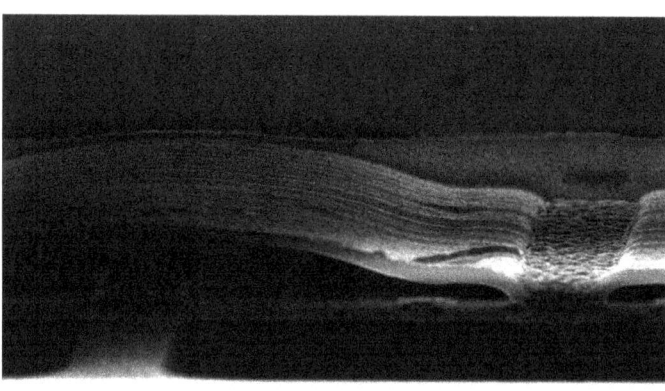

Figure 3. Ten-finger air-bridge device SEM.

3. Results

We fabricated a ten-finger common-source HEMT with an air-bridge and a gate length of $L_G = 4$ μm; the distance between the gate and source was $L_{GS} = 4$ μm, and the distance between the gate and drain was $L_{GD} = 12$ μm. In order to connect two adjacent sources (distance was 40 μm), a 40 μm span bridge with a source field plate structure was made through DWLGP.

The ten-finger devices connected with the air-bridge had good output performance. The measured output I-V characteristics of ten-finger common-source HEMT with air-bridge: V_{GS} was kept ordinally from 0 to 8 V (step 2 V), V_{DS} was swept from 0 to 20 V, and the source was tied to the ground. Their I-V characteristics were shown in Figure 4a; the saturation current of the devices was 240 mA/mm, and on-resistance was 36 Ω·mm. The measured threshold voltage (V_{TH}): V_{DS} was kept at 10 V, V_{GS} was swept from 0 to 5 V, and the source was tied to the ground. When it measured I_D at 1 mA/mm, the corresponding V_{GS} was the threshold voltage of the device. $V_{TH} = 1.8$ V (measured at $I_D = 1$ mA/mm) was shown in Figure 4b. The success of the ten-finger device was dependent on the stability of grayscale exposure, which the ten-finger device adopted a continuous air-bridge method to connect the source. It performed that fabricating the air-bridge through DWLGP had an excellent success rate. Up to now, without a bridge collapse after metal take-off, it indicated that the designed bridge structure had excellent reliability. It is noteworthy that when adopting the DWLGP method to mold the bridge, we need to control the exposure dose; too large of an exposure dose may expose the gate and directly connect the source to the grid after evaporating the metal, so the dose needs control within 10 mj. In order to ensure the pier line-width, it is necessary to use a laser scanning confocal microscope to view the photoresist morphology characteristics after the photoresist grayscale is exposed and developed. Incidentally, this laser scanning confocal microscope did not negatively affect the 3D photoresist structure.

Shock ionization and electron penetration are known to be the two most well-known physical mechanisms for the breakdown of HEMT devices [23,24]. There was a large peak at the gate electrode edge, especially if the gate was near the drain edge. If this large electric peak is not treated properly, it will cause electron penetration and the early breakdown of the device. In the design of a large-span bridge, the oblique floating-source field-plate structure is cleverly designed at the gate near the drain. It helps us to alleviate the electric field peak at the gate edge, which leads to significant breakdown voltage improvement. It can be seen from Figure 4c that the air-bridge floating-source field-plate structure HEMT had a high breakdown voltage: $V_{BR} = 715$ V; the breakdown voltage measurements were performed using the Agilent B1505A semiconductor parameter analyzer. For the breakdown voltage tests, we set the maximum voltage and current limit of the test at 1200 V and 1 mA, V_{GS} was set at 0 V, the source terminal was grounded, and V_D increased from 0 V (step = 5 V). During the measurements, the devices were covered with Fluorinert.

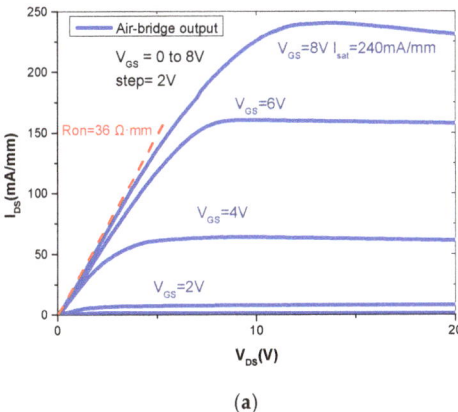

(a)

Figure 4. Cont.

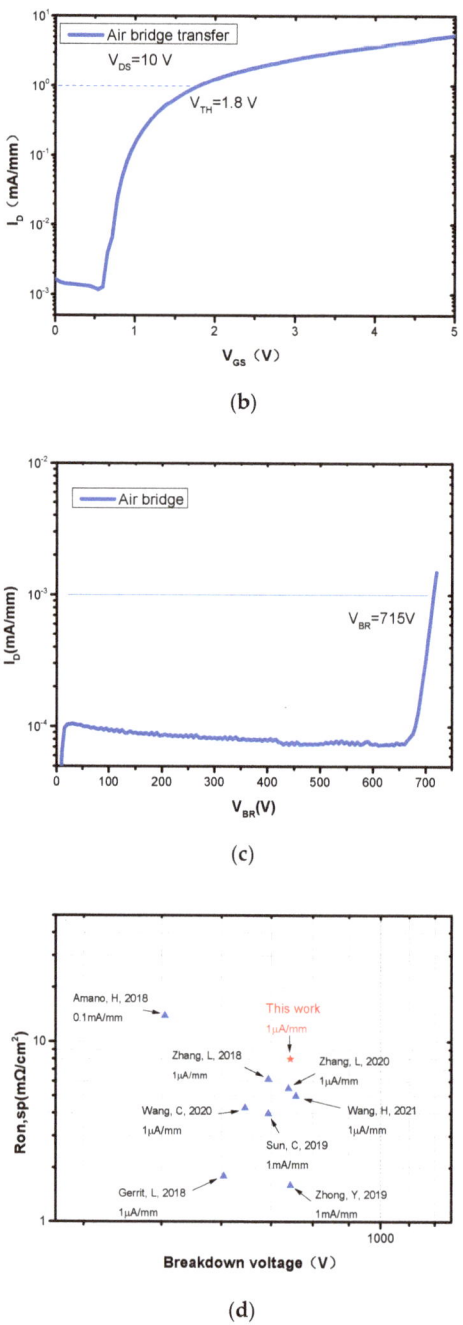

Figure 4. (**a**) ten-finger air-bridge normally-off I-V output, V_{GS} from 0 to 8 V (step 2 V), and V_{DS} from 0 to 20 V; (**b**) measured DC I-V transfer characteristics of ten-finger air-bridge normally-off structure. $V_{DS} = 10$ V; (**c**) Measured off-state characteristics of the fabricated ten-finger common-source HEMT breakdown voltage; (**d**) Ten-finger common-source HEMT V_{BR} and $R_{on,sp}$ comparisons with other p-GaN HEMT works [25–32].

Figure 4d shows the proposed device against other p-GaN HEMT on the Si substrate reported [25–32]. The high-specific R_{on} ($R_{on,sp}$) of our device was attributed to the excessive etching of the source/drain of the AlGaN layer before metal evaporation, leading to the increase in ohmic contact resistance. This could be improved by optimizing etching time in the future. The competitive V_{BR} could be attributed to the effective reduction in the high electric field crowded near the gate by using an air-bridge field plate. Compared to the AlGaN/GaN device (V_{BR} = 375 V) with an air-bridge field plate [5], the fabrication of the ten-finger air-bridge connection of p-GaN devices has a higher breakdown performance that may be attributed to the longer distance between the source and drain.

4. Conclusions

We confirmed the feasibility of the common-source ten-finger device with an air-bridge prepared using DWLGP. This common-source ten-finger device has excellent output and transfer performance, and the use of the common source improves the utilization rate of the wafer. The use of a float-source field-plate structure can reduce the peak electric field at the gate edge and obtain a good breakdown voltage. The use of DWLGP-prepared air-bridges is suitable for the fast iteration of a prototype.

Author Contributions: Conceptualization, Y.Z. and G.D.; methodology, P.Y. and Y.Z.; software, F.W.; validation, Y.Z., C.Y. and J.H.; formal analysis, F.W. and X.W.; investigation, Y.Z. and Q.F.; resources, M.H., Y.W. and X.W.; data Y.Z. and F.W.; writing—original draft preparation Y.Z.; supervision, X.W., W.C., H.J., H.C. and Y.W.; project administration, Y.W. and W.X. All authors have read and agreed to the published version of the manuscript.

Funding: This work was supported in part by the Key-Area Research and Development Program of Guangdong Province, China, under Grant No.2020B010174001, and in part by the Guang-dong Basic and Applied Basic Research Foundation, China, under Grant No.2020A1515110567.

Data Availability Statement: The data presented in this study are available on request from the corresponding author.

Conflicts of Interest: The authors declare no conflict of interest.

References

1. Chen, K.J. GaN-on-Si Power Technology: Devices and Applications. *IEEE Trans. Electron Device* **2017**, *64*, 779–795. [CrossRef]
2. Efthymiou, L. On the physical operation and optimization of the p-GaN gate in normally-off GaN HEMT devices. *Appl. Phys. Lett.* **2017**, *110*, 123502. [CrossRef]
3. Zheng, Z. Enhancement-Mode GaN p-Channel MOSFETs for Power Integration. In Proceedings of the 2020 32nd International Symposium on Power Semiconductor Devices and ICs (ISPSD), Vienna, Austria, 13–18 September 2020; pp. 525–528.
4. Zhou, F. 1.2 kV/25 A Normally off P-N Junction/AlGaN/GaN HEMTs With Nanosecond Switching Characteristics and Robust Overvoltage Capability. *IEEE Trans. Power Electron.* **2022**, *37*, 26–30. [CrossRef]
5. Xie, G. Breakdown-Voltage-Enhancement Technique for RF-Based AlGaN/GaN HEMTs With a Source-Connected Air-Bridge Field Plate. *IEEE Electron Device Lett.* **2012**, *33*, 670–672. [CrossRef]
6. Dora, Y.; Chakraborty, A.; Mccarthy, L.; Keller, S.; Denbaars, S.P.; Mishra, U.K. High breakdown voltage achieved on AlGaN/GaN HEMTs with integrated slant field plates. *IEEE Electron Device Lett.* **2006**, *27*, 713–715. [CrossRef]
7. Coffie, R. Slant field plate model for field-effect transistors. *IEEE Trans. Electron Device* **2014**, *61*, 2867–2872. [CrossRef]
8. Kirchner, R. Bio-inspired 3D funnel structures made by grayscale electron-beam patterning and selective topography equilibration. *Microelectron. Engineer.* **2015**, *141*, 107–111. [CrossRef]
9. Kim, J. Controlling resist thickness and etch depth for fabrication of 3D structures in electron-beam grayscale lithography. *Microelectron. Engineer.* **2007**, *84*, 2859–2864. [CrossRef]
10. Totsu, K. Fabrication of three-dimensional microstructure using maskless gray-scale lithography. *Sens. Actuators A Phys.* **2006**, *130–131*, 387–392. [CrossRef]
11. Ngo, T.H.; Comyn, R.; Chenot, S.; Brault, J.; Damilano, B.; Vezian, S.; Frayssinet, E.; Cozette, F.; Defrance, N.; Lecourt, F.; et al. Combination of selective area sublimation of p-GaN and regrowth of AlGaN for the co-integration of enhancement mode and depletion mode high electron mobility transistors. *Solid-State Electron.* **2022**, *188*, 108210. [CrossRef]
12. Greco, G.; Iucolano, F.; Roccaforte, F. Review of technology for normally-off HEMTs with p-GaN gate. *Mater. Sci. Semicond. Process.* **2018**, *78*, 96–106. [CrossRef]

13. Green, R.T.; Luxmoore, I.J.; Lee, K.B.; Houston, P.A.; Ranalli, F.; Wang, T.; Parbrook, P.J.; Uren, M.J.; Wallis, D.J.; Martin, T. Characterization of gate recessed GaN/AlGaN/GaN high electron mobility transistors fabricated using a $SiCl_4/SF_6$ dry etch recipe. *J. Appl. Phys.* **2010**, *108*, 013711. [CrossRef]
14. Chang, Y.C.; Ho, Y.L.; Huang, T.Y.; Huang, D.W.; Wu, C.H. Investigation of Normally-Off p-GaN/AlGaN/GaN HEMTs Using a Self-Terminating Etching Technique with Multi-Finger Architecture Modulation for High Power Application. *Micromachines* **2021**, *12*, 432. [CrossRef]
15. Taube, A.; Kamiński, M.; Ekielski, M.; Kruszka, R.; Jankowska-Śliwińska, J.; Michałowski, P.P.; Zdunek, J.; Szerling, A. Selective etching of p-GaN over $Al_{0.25}Ga_{0.75}N$ in $Cl_2/Ar/O_2$ ICP plasma for fabrication of normally-off GaN HEMTs. *Mater. Sci. Semicond. Process.* **2021**, *122*, 105450. [CrossRef]
16. Zhu, Y.; Huang, R.; Li, Z.; Hao, H.; An, Y.; Liu, T.; Zhao, Y.; Shen, Y.; Guo, Y.; Li, F.; et al. Interface analysis of TiN/n-GaN ohmic contacts with high thermal stability. *Appl. Surf. Sci.* **2019**, *481*, 1148–1153. [CrossRef]
17. Wang, L.; Kim, D.H.; Adesida, I. Direct contact mechanism of Ohmic metallization to AlGaN/GaN heterostructures via Ohmic area recess etching. *Appl. Phys. Lett.* **2009**, *95*, 172107. [CrossRef]
18. Vetury, R.; Zhang, N.Q.; Keller, S.; Mishra, U.K. Electrical and Computer Engineering Department, University of California, Santa Barbara, CA, USA The impact of surface states on the DC and RF characteristics of AlGaN/GaN HFETs. *IEEE Trans. Electron Device* **2001**, *48*, 560–566. [CrossRef]
19. Geng, K.; Chen, D.; Zhou, Q.; Wang, H. AlGaN/GaN MIS-HEMT with PECVD SiNx, SiON, SiO_2 as gate dielectric and passivation layer. *Electronics* **2018**, *7*, 416. [CrossRef]
20. Xu, D.; Chu, K.; Diaz, J.; Zhu, W.; Roy, R.; Seekell, P.; Pleasant, L.M.; Isaak, R.; Yang, X.; Nichols, K.; et al. Performance enhancement of GaN high electron-mobility transistors with atomic layer deposition Al_2O_3 passivation. In Proceedings of the 2012 Lester Eastman Conference on High Performance Devices (LEC), Providence, RI, USA, 7–9 August 2012; IEEE: Piscataway, NJ, USA, 2012; pp. 1–3.
21. Bae, C.; Lucovsky, G. Low-temperature preparation of Ga $N-SiO_2$ interfaces with low defect density. I. Two-step remote plasma-assisted oxidation-deposition process. *J. Vac. Sci. Technol. A Vac. Surf. Film.* **2004**, *22*, 2402–2410. [CrossRef]
22. Lu, X.; Jiang, H.; Liu, C.; Zou, X.; Lau, K.M. Off-state leakage current reduction in AlGaN/GaN high electron mobility transistors by combining surface treatment and post-gate annealing. *Semicond. Sci. Technol.* **2016**, *31*, 055019. [CrossRef]
23. Uren, M.J. Punch-through in short-channel AlGaN/GaN HFETs. *IEEE Trans. Electron Devices* **2006**, *53*, 395–398. [CrossRef]
24. Bahat-Treidel, E. Punchthrough-Voltage Enhancement of AlGaN/GaN HEMTs Using AlGaN Double-Heterojunction Confinement. *IEEE Trans. Electron Devices.* **2008**, *55*, 3354–3359. [CrossRef]
25. Zhong, Y. Normally-off HEMTs With Regrown p-GaN Gate and Low-Pressure Chemical Vapor Deposition SiNx Passivation by Using an AlN Pre-Layer. *IEEE Electron Device Lett.* **2019**, *40*, 1495–1498. [CrossRef]
26. Sun, C. Normally-off p-GaN/AlGaN/GaN high-electron-mobility transistors using oxygen plasma treatment. *Appl. Phys. Express.* **2019**, *12*, 051001. [CrossRef]
27. Amano, H. The 2018 GaN power electronics roadmap. *J. Phys. D Appl. Phys.* **2018**, *51*, 163001. [CrossRef]
28. Gerrit, L. Self-Aligned Process for Selectively Etched p-GaN-Gated AlGaN/GaN-on-Si HFETs. *IEEE Trans. Electron Devices* **2018**, *65*, 3732–3738.
29. Li, Z. P-GaN Gate Power Transistor With Distributed Built-in Schottky Barrier Diode for Low-loss Reverse Conduction. *IEEE Electron Device Lett.* **2020**, *41*, 341–344.
30. Li, Z. AlGaN-Channel Gate Injection Transistor on Silicon Substrate With Adjustable 4–7-V Threshold Voltage and 1.3-kV Breakdown Voltage. *IEEE Electron Device Lett.* **2018**, *39*, 1026–1029.
31. Wang, C. E-Mode p-n Junction/AlGaN/GaN (PNJ) HEMTs. *IEEE Electron Device Lette.* **2020**, *41*, 545–548. [CrossRef]
32. Wang, H. High-performance reverse blocking p-GaN HEMTs with recessed Schottky and p-GaN isolation blocks drain. *Appl. Phys. Lett.* **2021**, *119*, 023507. [CrossRef]

Disclaimer/Publisher's Note: The statements, opinions and data contained in all publications are solely those of the individual author(s) and contributor(s) and not of MDPI and/or the editor(s). MDPI and/or the editor(s) disclaim responsibility for any injury to people or property resulting from any ideas, methods, instructions or products referred to in the content.

Article

Analysis of Photo-Generated Carrier Escape in Multiple Quantum Wells

Jiaping Guo [1], Weiye Liu [1], Ding Ding [1], Xinhui Tan [1], Wei Zhang [1], Lili Han [1], Zhaowei Wang [1], Weihua Gong [1], Jiyun Li [1], Ruizhan Zhai [1], Zhongqing Jia [1], Ziguang Ma [2], Chunhua Du [3], Haiqiang Jia [3] and Xiansheng Tang [1,*]

- [1] Laser Institute, Qilu University of Technology (Shandong Academy of Sciences), Jinan 250014, China
- [2] Huawei Technologies Co., Ltd., Beijing 100095, China
- [3] Institute of Physics, Chinese Academy of Sciences, Beijing 100083, China
- * Correspondence: 18811681359@163.com; Tel.: +86-188-1168-1359

Abstract: Recent experiments have shown that more than 85% of photo-generated carriers can escape from multiple quantum wells (MQWs) sandwiched between p-type and n-type layers (PIN). In this work, we quantitatively analyze the relationship between the energy of carriers and the height of potential barriers to be crossed, based on the GaAs/InGaAs quantum well structure system, combined with the Heisenberg uncertainty principle. It was found that that the energy obtained by electrons from photons is just enough for them to escape, and it was found that the energy obtained by the hole is just enough for it to escape due to the extra energy calculated, based on the uncertainty principle. This extra energy is considered to come from photo-generated thermal energy. The differential reflection spectrum of the structure is then measured by pump–probe technology to verify the assumption. The experiment shows that the photo-generated carrier has a longer lifetime in its short circuit (SC) state, and thus it possesses a lower structure temperature than that in open circuit (OC). This can only explain a thermal energy reduction caused by the continuous carrier escape in SC state, indicating an extra thermal energy transferred to the escaping carriers. This study is of great significance to the design of new optoelectronic devices and can improve the theory of photo-generated carrier transports.

Keywords: photo-generated carriers; escape; uncertainty principle

1. Introduction

Semiconductor materials are the cornerstones of the information society, and the progress of science and technology cannot be separated from the development of semiconductor materials and processes. In addition to the important applications in the field of electronics, semiconductor materials also have high application potential in the field of optoelectronics. With the advancement of semiconductor theory and manufacturing technology, quantum well (QW) and super-lattice structures have been developed. The quantum well structure has played an important role in the field of electricity-to-light conversion since its invention [1–9]. Light-emitting diodes (LEDs) [10–12] and lasers [13–15] both adopt multi-quantum well (MQW) structures due to their strong carrier capture and confinement capabilities in the electricity-to-light conversion field. These two devices achieved good commercial applications, and they are gradually changing and enriching people's lives. As a new generation of lighting sources, LEDs have been gradually replacing incandescent light sources. The laser is more widely used in industrial, military, and other fields. However, as for the light-to-electricity conversion, the use of low-dimensional structures is somewhat limited. Quantum well infrared detectors (QWIPs) are the most widely used structures [16–18], which mainly exploit transitions between sub-bands. However, MQWs are hard to apply in light-to-electricity conversion devices based on the inter-band transition because, according to the classical theory, the photo-generated carriers would relax to the ground state of the low-dimensional materials, from where the photo-generated carriers cannot escape to form a photocurrent due to quantum confinement [19]. Due to quantum confinement, the carriers in quantum wells can only disappear through relaxation and recombination and cannot escape. Therefore, the application of low-dimensional semiconductor structure in the field of photoelectric conversion is limited.

Recently, many research groups have found that most part of photo-generated carriers would escape from MQWs when the MQWs' structures are placed in the depletion region of the PN junction (PIN structure) [20–23]. By measuring the photoluminescence (PL) spectra in the case of open circuits (OCs) and short circuits (SCs), respectively, it was found that there is an obvious fluorescence quenching in the case of SC, accompanied by obvious photocurrent production [24–28]. This new phenomenon breaks our understanding of the carrier limiting effect of traditional low-dimensional structures. This shows that the photo-generated carriers can escape efficiently and can be used in photo-electric conversion devices. Furthermore, this phenomenon is not special and can be found in many material systems. We have investigated a large number of low dimensional material systems, including GaN/InGaN quantum well structures [21], GaAs/InGaAs quantum well structures [20], and GaAs/InAs quantum dot (QD) structures [29], all of which indicate that, when the thickness of the MQWs structure region is less than 100 nm, more than 85% of carriers can escape by comparing their integral strength of the PL peak under the OC and SC states. This special structure can be used to make new types of optoelectronic devices, such as quantum well solar cells and photodetectors (IQWIP) [25,30], which widen the range of applications of quantum wells. Meanwhile, as for the NIN structure that the MQW structures are placed into in the depletion region of the NN junction (NIN structure), there is no obvious fluorescence quenching phenomenon in the case of SC, even if bias voltage is applied [21,22,31]. Furthermore, this large escape rate cannot be explained by tunneling theory or thermionic emission theory. These theories can only explain some special structures in the special cases, but they cannot cover all cases. In some cases, a combination of the two theories is needed to explain a particular situation. Therefore, these two theories are not complete and need new interpretation. In this paper, we quantitatively analyze the relationship between the energy of carriers and the height of potential barriers to be crossed, based on the GaAs/InGaAs MQWs structure system of ref. [20], combined with the Heisenberg uncertainty principle [32]. We compared the energy difference between the energy level and the energy obtained by the excited carriers. The energy is analyzed from a quantitative point of view, and, combined with the energy introduced by the uncertainty principle, further numerical comparison is made. Based on the above analysis, some energy sources are guessed, and the pump–probe technique was applied to verify our relevant hypotheses. This study is of great significance to the design of new optoelectronic devices and can improve the theory of photo-generated carrier transport.

2. Analysis and Discussion

The PIN structure investigated in this work is based on GaAs/InGaAs MQWs structure [20]. Its epitaxial structure and energy band structure are shown in Figure 1a. The structure includes 10 quantum wells of InGaAs with thickness of 5 nm, and the composition of indium is 0.2. The width of barrier of GaAs is 20 nm. Figure 1 shows the energy band diagrams of PIN structure, the corresponding MQWs structure, and the sub-band energies, respectively. It can be seen that the conduction band level of GaAs is 0.6242 eV, and the valence band level is -0.7951 eV, showing that the band gap is 1.42 eV, which is consistent with the classical data of GaAs at 300 K. From Figure 1b, we can see that the conduction band level of In(0.2)GaAs is 0.4765 eV, and the valence band level is -0.7115 eV; for the electron, the first confined energy level in the well is 0.5424 eV, and, for the hole, the first confined energy level in the well is -0.7264 eV. This shows a band gap of 1.27 eV between two confined energies, corresponding to an emission wavelength of 976 nm, which is almost in agreement with that reported in [20]. Here, the strain is not considered in the calculation process, and the thickness of samples grown in the experiment is associated with a fluctuation, which results in a 6 nm wavelength fluctuation. So, this calculation is acceptable.

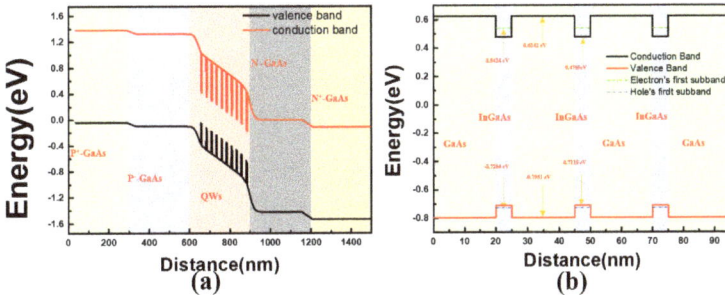

Figure 1. (a) Band diagram of the PIN structure. (b) Band diagram of ordinary quantum well structures.

According to the above data, we can find that the energy difference between the electron's first confined energy level and the energy of the barrier level of the electron is 81.8 meV, and the energy difference between the hole's first confined energy level and the energy of the barrier level of the hole is 68.7 meV. The energy difference between the energy of barrier level of the electron and the hole's first confined energy level is 1.35 eV, which corresponds to an excitation wavelength of 918 nm. When an excitation wavelength of 915 nm is used [20], the photon energy is greater than 1.355 eV. Therefore, the excited electrons with enough energy can escape directly. Additionally, electrons are not localized in quantum wells. So, we only need to consider the energy of excited holes.

The hole could also gain energy through the scattering from carriers and lattice, though the whole energy of the incident photon is carried away by the electron through the transition. The average thermal equilibrium energy of a large number of particles can be obtained by equilibrium theory. Additionally, the energy per free dimension is 1/2 kT, where k represents the Boltzmann constant, and T represents the absolute temperature. The carrier only has two free dimensions in the confined energy level. So, the energy of the hole obtained from carriers or lattices by scattering is kT. However, the above relevant experiments have shown that confined carriers can escape from MQWs in the PIN structure, which showed the properties of free carriers in three dimensions. Therefore, the hole would have three free dimensions, and the resulting energy is 3/2 kT, which is 39 meV at room temperature.

It should be noted that there Is a strong built-in electric field in the depletion region of the PN junction, which will accelerate the hole to obtain a certain velocity. The simulated distribution of the electric field in the structure is shown in Figure 2a. It can be seen that the electric field intensity in the MQW region is greater than 2×10^4 V/cm. From Figure 2b, we can see the relationship between the carrier saturation drift velocity and the electric field intensity in the GaAs material system is based on equations. The saturation drift velocity of holes increases with the increase in electric field intensity, eventually approaching to 10^7 cm/s. There exists velocity overshoot due to very short acceleration time of the hole in the MQWs. Meanwhile, due to the separation energy level in the quantum wells, the scattering effect of the hole will be reduced, so its saturation velocity will also increase. We treated the saturation velocity of the cavity in the well as 10^7 cm/s. Therefore, the kinetic energy obtained by the hole from the electric field is $1/2\ mv^2$, corresponding to 17 meV. Combined with the above 39 meV thermal energy, the current total energy obtained is 56 meV, which is still less than the 68.7 meV that needs to be crossed. So, we need to think about other forms of energy.

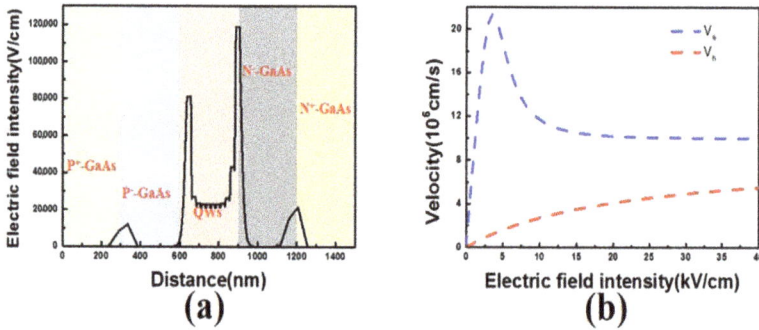

Figure 2. (a) Electric field intensity distribution in the PIN structure. (b) Variation of carrier saturation drift velocity vs. electric field intensity in GaAs.

Since holes also have quantum properties, we consider the energy due to uncertainty relations. Finally, according to the Heisenberg uncertainty principle $\Delta E \Delta t \geq \hbar/2$, where ΔE represents the difference between the measured energy and the actual energy, and Δt, the difference between the measured time and the actual time, $\hbar$ reduced Planck's constant. The width of the well is known to be 5 nm, and the velocity of the hole is 10^7 cm/s, giving an average escape time of the hole of 25 fs. It has been confirmed, from previous experiments, that the escape time of photo-generated carriers is on the order of femtoseconds to picoseconds. This is in good agreement with the reported data [31], and it is accurate in terms of magnitude. Now that the time is accurate, the energy uncertainty ΔE can be estimated by substituting the time and the reduced Planck's constant, which is ~13 meV.

From the above analysis, we can see that the energy of the hole contains thermal energy (39 meV), kinetic energy (17 meV), and uncertain energy (13 meV). Therefore, the hole in the quantum wells in the PIN structure has an energy of 69 meV, which is just larger than the barrier potential of 68.7 meV. This implies that, after considering the energy introduced by the uncertainty principle, the hole can escape from the MQWs. So, both holes and electrons can escape from the PIN structure and enter the external circuit, which results in fluorescence quenching in the case of SC. Additionally, here, we have a quantitative explanation of why photo-generated carriers can escape to MQWs.

For the state of OC, both sides of the wafer accumulate different carriers, which would introduce a new electric field, whose direction is different than that of the built-in electric field of the PN junction, as shown in Figure 3a, making the hole lose part of the energy source.

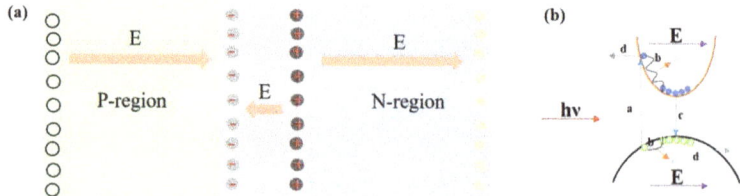

Figure 3. (a) The electric field inside the structure under OC. (b) Four processes that photo-generated carriers go through.

Since the electric field formed by the accumulation of charge carriers on both sides of the epitaxial structure is not damaged by the outside cases, and the surface–composite charge carriers are constantly supplemented by the internal photo-generated carriers, this maintains the stability of the electric field. The holes cannot obtain enough energy for continuous escape, which makes the fluorescence intensity still strong in the OC state, even in the case of escape. From Figure 3b, we can see that there are four processes for photo-generated carriers in the PIN structure: generation, relaxation, recombination, and escape. Although the time for photo-generated carriers to escape from a quantum well is on the order of femtoseconds, it is on the order of picoseconds for the MQW region, with a width of hundreds of nanometers or even microns. The photo-generated carriers either escape or undergo radiative recombination. It has been proven that the carrier escapes first [28]. The fluorescence intensity is consistent with that of OC when a bias voltage equivalent to the built-in electric field is applied. This fully indicates that the existence of escaping carriers reduces the built-in electric field. So, photogenic carriers cannot escape continuously.

According to the conservation of energy, the uncertain energy obtained by the hole should also be converted from another energy source. It is preliminarily speculated that this part of the energy comes from photon energy that fails to generate carriers by excitation, that is, heat energy. Next, we designed an experiment to verify this hypothesis.

It is well known that the lifetime of carriers decreases with temperature increase. Therefore, we can compare the lifetime of the carriers to explore the local surface temperature of the sample. Photo-generated carriers are a kind of non-equilibrium carrier. Their behavior follows a certain statistical theory. In general, the decay is exponential. The concentration of photo-generated carriers will affect the reflectance of the sample surface. Therefore, the carrier lifetime can be detected by measuring the differential reflection spectrum, based on the pump–probe technique. The pump–probe technique was first proposed by Toepler. Two femtosecond (fs) pulses, with time delay, are used, in which the one with higher energy and earlier time is used as the pump light, while the one with lower energy and later time is used as the probe light to excite and probe the samples, respectively. The pump light and the probe light are obtained by the same femtosecond laser beam passing through the beam splitter mirror, with one beam of high energy as the pump light and the other beam of low energy as the probe light. After the probe light passes through a displacement platform, there will be an optical path difference between the probe light and the pump light, and then a certain time interval will be generated. This is performed so that they can reach the surface of the sample successively. The pump light excites the sample to the excited state, and the probe light with time delay arrives later. The probe sample evolves with time after being excited. Figure 4a shows the schematic diagram of the pump–probe technology. The laser used has a wavelength of 808 nm with 1 kHz repetition frequency and 1 mW energy. The pulsed laser beam passed through a polarization beam splitter (PBS), which was divided into two pulsed beams, pump and probe, with an energy ratio of 10:1. The pump light shines on the sample, which is consistent with ref. [20], after passing through a corner reflector, which is mounted on a step motor. Therefore, the optical path difference between the pump light and the probe light, i.e., the time difference between the two pulsed light beams arriving at the sample surface, can be adjusted. The surface reflectivity of the sample is measured by using probe light with and without pump excitation, respectively, so that the differential reflection spectrum is obtained. The photon energy, corresponding to 808 nm wavelength, is larger than the band gap of GaAs and the band gap of InGaAs in the quantum well, so the photo-generated carriers can be excited simultaneously in the quantum well and on the surface. By studying the dynamic behavior of surface photo-generated carriers, the lifetime and local temperature variation are determined. The same measurement was carried out while the sample was further heated with a 532 nm laser for the purpose of enhancing the measurement contrast. By setting the sample to OC state and SC state with and without laser heating, we obtain the measurement results, which are shown in Figure 4b, which shows the lifetime of photo-generated carriers under different condition.

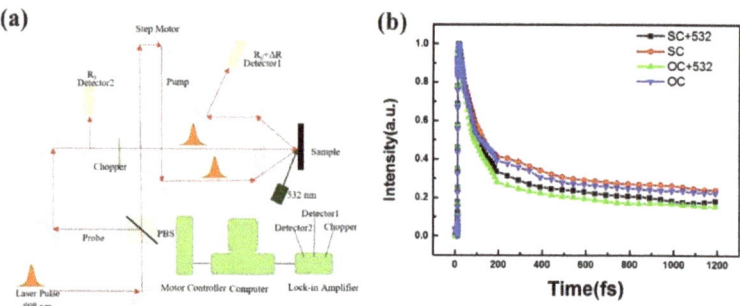

Figure 4. (**a**) Physical diagram of the optical path of pump-probe technology. (**b**) Differential reflection spectra of samples in different states.

From Figure 4b, it can be seen that the lifetime of the SC state is longer than that of the OC state, showing that the sample temperature of SC state is lower than that of the OC state. It has been reported that, under the condition of SC, most photo-excited carriers escape from the quantum wells to generate photo-excited current, rather than relax to the ground state of quantum wells and recombine to emit light [20]. Since holes need energy to escape in the SC state, the heat energy may provide this excess energy, giving rise to a lower sample temperature than that in OC state. In the OC state, the sample temperature is higher due to no continuous escape phenomena.

When 532 nm laser heating is applied, both lifetimes of photo-generated carriers under OC and SC conditions become smaller, but they show the same trend, indicating a higher sample temperature induced by light heating. In this case, however, the lifetime of the SC state still is longer than that of the OC state. From the lifetime of the carrier in different states, one can deduce that the heat generated by light makes a contribution to the carrier escape in the SC state, resulting in a lower sample temperature than that in the OC state. This implies that heat generated by light may provide excess energy for carrier escape in the SC state.

3. Conclusions

In this paper, we mainly analyzed the energy of the hole from multiple aspects. In terms of this specific structure and the photo excitation source used (915 nm), we can know that the photo-generated electron can escape from quantum wells with enough energy. The energy uncertainty of the hole is calculated (13 meV), based on the uncertainty principle. Combined with thermal energy (39 meV) and kinetic energy (17 meV), we can find that the energy of the hole is 69 meV, which is greater than the barrier potential of 68.7 meV. This implies that the hole has enough energy to escape. We have assumed that the excess energy comes from the photon that has not excited the carriers, that is, photo-generated heat energy, and we verified this by using pump–probe technology. It has been found that, in the SC state, energy is required for carriers to escape, and this excess energy comes from light heating energy, so the sample temperature in this state is lower than that of the OC state, and the lifetime of carriers is longer than that of the OC state. In the OC state, the sample temperature is higher because of the lack of continuous escape phenomenon, and the lifetime of carriers is shorter. The conclusion of this work would help better understand the design of new solar cells, photodetectors, and other photoelectric devices, which are based on MQWs. It also provides a complement to the transport theory of photo-generated carriers.

Author Contributions: Conceptualization, R.Z., Z.J., Z.M., C.D., H.J. and X.T. (Xiansheng Tang); methodology, Z.M., C.D., H.J. and X.T. (Xiansheng Tang); software, X.T. (Xiansheng Tang); validation, W.Z., L.H., Z.W., W.G. and J.L.; formal analysis, X.T. (Xiansheng Tang); investigation, X.T. (Xiansheng Tang); resources, X.T. (Xiansheng Tang), W.Z., L.H. and Z.W.; data curation, X.T. (Xiansheng Tang); writing—original draft preparation, J.G., W.L., D.D. and X.T. (Xinhui Tan); writing—review and editing, X.T. (Xiansheng Tang). All authors have read and agreed to the published version of the manuscript.

Funding: This work was supported by National Key research and development Program (Grant No. 2021YFB3201904). This work was supported by National Natural Science Foundation of China (No. 62005138). This work was supported by Qilu University of Technology (Shandong Academy of Sciences) Peixin fund project (Grant No. 2022PX080). This work was supported by Qilu University of Technology (Shandong Academy of Sciences) International Cooperation Projects (Grant No. 2022GH001). This work was also supported by Qilu University of Technology (Shandong Academy of Sciences) Computer Science and Technology "Four Plans" talent introduction and Multiplication plan project (Grant No. 2021YY01002) and supported by the Youth fund of the Shandong Natural Science Foundation (Grant Nos. ZR2020QF098 and ZR2022QF115). This work was also supported by Project of Jinan (Grant No. 2020GXRC032).

Data Availability Statement: Data are available upon request from the authors.

Conflicts of Interest: The authors declare no conflict of interest.

References

1. Bushnell, D.B.; Tibbits, T.N.D.; Barnham, K.W.J.; Connolly, J.P.; Mazzer, M.; Ekins-Daukes, N.J.; Roberts, J.S.; Hill, G.; Airey, R. Effect of well number on the performance of quantum-well solar cells. *J. Appl. Phys.* **2005**, *97*, 124908. [CrossRef]
2. Courel, M.; Rimada, J.C.; Hernández, L. GaAs/GaInNAs quantum well and superlattice solar cell. *Appl. Phys. Lett.* **2012**, *100*, 073508. [CrossRef]
3. Wang, S.; Long, H.; Zhang, Y.; Chen, Q.; Dai, J.; Zhang, S.; Chen, J.; Liang, R.; Xu, L.; Wu, F.; et al. Monolithic integration of deep ultraviolet LED with a multiplicative photoelectric converter. *Nano Energy* **2019**, *66*, 104181. [CrossRef]
4. Elahi, E.; Suleman, M.; Nisar, S.; Sharma, P.R.; Iqbal, M.W.; Patil, S.A.; Kim, H.; Abbas, S.; Chavan, V.D.; Dastgeer, G.; et al. Robust approach towards wearable power efficient transistors with low subthreshold swing. *Mater. Today Phys.* **2022**, *30*, 100943. [CrossRef]
5. Dastgeer, G.; Nisar, S.; Shahzad, Z.M.; Rasheed, A.; Kim, D.K.; Jaffery, S.H.A.; Eom, J. Low-Power Negative-Differential-Resistance Device for Sensing the Selective Protein via Supporter Molecule Engineering. *Adv. Sci.* **2023**, *10*, 2204779. [CrossRef] [PubMed]
6. Dastgeer, G.; Shahzad, Z.M.; Chae, H.; Kim, Y.H.; Ko, B.M.; Eom, J. Bipolar Junction Transistor Exhibiting Excellent Output Characteristics with a Prompt Response against the Selective Protein. *Adv. Funct. Mater.* **2022**, *32*, 2204781. [CrossRef]
7. Dastgeer, G.; Afzal, A.M.; Aziz, J.; Hussain, S.; Jaffery SH, A.; Kim, D.K.; Assiri, M.A. Flexible memory device composed of metal-oxide and two-dimensional material (SnO2/WTe2) exhibiting stable resistive switching. *Materials* **2021**, *14*, 7535. [CrossRef] [PubMed]
8. Dastgeer, G.; Afzal, A.M.; Jaffery, S.H.A.; Imran, M.; Assiri, M.A.; Nisar, S. Gate modulation of the spin current in graphene/WSe2 van der Waals heterostructure at room temperature. *J. Alloys Compd.* **2022**, *919*, 165815. [CrossRef]
9. Dastgeer, G.; Afzal, A.M.; Nazir, G.; Sarwar, N. p-GeSe/n-ReS2 heterojunction rectifier exhibiting a fast photoresponse with ultra-high frequency-switching applications. *Adv. Mater. Interfaces* **2021**, *8*, 2100705. [CrossRef]
10. Yeh, D.-M.; Huang, C.-F.; Chen, C.-Y.; Lu, Y.-C.; Yang, C.C. Surface plasmon coupling effect in an InGaN/GaN single-quantum-well light-emitting diode. *Appl. Phys. Lett.* **2007**, *91*, 171103. [CrossRef]
11. Yadav, G.; Dewan, A.; Tomar, M. Electroluminescence study of InGaN/GaN QW based p-i-n and inverted p-i-n junction based short-wavelength LED device using laser MBE technique. *Opt. Mater.* **2022**, *126*, 112149. [CrossRef]
12. Ahmad, S.; Raushan, M.; Kumar, S.; Dalela, S.; Siddiqui, M.; Alvi, P. Modeling and simulation of GaN based QW LED for UV emission. *Optik* **2018**, *158*, 1334–1341. [CrossRef]
13. Hansen, M.; Piprek, J.; Pattison, P.M.; Speck, J.S.; Nakamura, S.; DenBaars, S.P. Higher efficiency InGaN laser diodes with an improved quantum well capping configuration. *Appl. Phys. Lett.* **2002**, *81*, 4275–4277. [CrossRef]
14. Tansu, N.; Mawst, L.J. Current injection efficiency of InGaAsN quantum-well lasers. *J. Appl. Phys.* **2005**, *97*, 054502. [CrossRef]
15. Margetis, J.; Zhou, Y.; Dou, W.; Grant, P.C.; Alharthi, B.; Du, W.; Wadsworth, A.; Guo, Q.; Tran, H.; Ojo, S.; et al. All group-IV Si-GeSn/GeSn/SiGeSn QW laser on Si operating up to 90 K. *Appl. Phys. Lett.* **2018**, *113*, 221104. [CrossRef]
16. Pan, J.L.; Fonstad, C.G. Theory, fabrication and characterization of quantum well infrared photodetectors. *Mater. Sci. Eng. R Rep.* **2000**, *28*, 65–147. [CrossRef]
17. Zhou, X.; Li, N.; Lu, W. Progress in quantum well and quantum cascade infrared photodetectors in SITP. *Chin. Phys. B* **2019**, *28*. [CrossRef]

18. Wu, W.; Bonakdar, A.; Mohseni, H. Plasmonic enhanced quantum well infrared photodetector with high detectivity. *Appl. Phys. Lett.* **2010**, *96*, 161107. [CrossRef]
19. Seeger, K. *Semiconductor Physics*; Springer Science & Business Media: Berlin/Heidelberg, Germany, 2013.
20. Sun, Q.; Wang, L.; Wang, Y.; Ma, Z.; Chen, H. Direct Observation of Carrier Transportation Process in InGaAs/GaAs Mul-tiple Quantum Wells Used for Solar Cells and Photodetectors. *Chin. Phys. Lett.* **2016**, *33*, 103–106. [CrossRef]
21. Wu, H.; Ma, Z.; Jiang, Y.; Wang, L.; Yang, H.; Li, Y.; Zuo, P.; Jia, H.; Wang, W.; Zhou, J.; et al. Direct observation of the carrier transport process in InGaN quantum wells with a pn-junction. *Chin. Phys. B* **2016**, *25*, 117803. [CrossRef]
22. Li, Y.; Jiang, Y.; Die, J.; Wang, C.; Yan, S.; Wu, H.; Ma, Z.; Wang, L.; Jia, H.; Wang, W.; et al. Visualizing light-to-electricity conversion process in InGaN/GaN multi-quantum wells with a p–n junction. *Chin. Phys. B* **2018**, *27*, 097104. [CrossRef]
23. Yang, H.; Ma, Z.; Jiang, Y.; Wu, H.; Zuo, P.; Zhao, B.; Jia, H.; Chen, H. The enhanced photo absorption and carrier transport tation of InGaN/GaN Quantum Wells for photodiode detector applications. *Sci. Rep.-UK* **2017**, *7*, 43357. [CrossRef]
24. Lim, S.; Ko, Y.; Cho, Y. A quantitative method for determination of carrier escape efficiency in GaN-based light-emitting diodes: A comparison of open- and short-circuit photoluminescence. *Appl. Phys. Lett.* **2014**, *104*, 91104. [CrossRef]
25. Watanabe, N.; Mitsuhara, M.; Yokoyama, H.; Liang, J.; Shigekawa, N. Influence of InGaN/GaN multiple quantum well structure on photovoltaic characteristics of solar cell. *Jpn. J. Appl. Phys.* **2014**, *53*, 112301. [CrossRef]
26. Lang, J.R.; Young, N.G.; Farrell, R.M.; Wu, Y.-R.; Speck, J.S. Carrier escape mechanism dependence on barrier thickness and temperature in InGaN quantum well solar cells. *Appl. Phys. Lett.* **2012**, *101*, 181105. [CrossRef]
27. Schubert, M.F.; Xu, J.; Dai, Q.; Mont, F.W.; Kim, J.K.; Schubert, E.F. On resonant optical excitation and carrier escape in GaInN/GaN quantum wells. *Appl. Phys. Lett.* **2009**, *94*, 081114. [CrossRef]
28. Song, J.-H.; Kim, H.-J.; Ahn, B.-J.; Dong, Y.; Hong, S.; Song, J.-H.; Moon, Y.; Yuh, H.-K.; Choi, S.-C.; Shee, S. Role of photovoltaic effects on characterizing emission properties of InGaN/GaN light emitting diodes. *Appl. Phys. Lett.* **2009**, *95*, 263503. [CrossRef]
29. Wang, W.; Wang, L.; Jiang, Y.; Ma, Z.; Sun, L.; Liu, J.; Sun, Q.; Zhao, B.; Wang, W.; Liu, W.; et al. Carrier transport in III–V quantum-dot structures for solar cells or photodetectors. *Chin. Phys. B* **2016**, *25*, 097307. [CrossRef]
30. Zou, Y.; Honsberg, C.B.; Freundlich, A.; Goodnick, S.M. Simulation of Electron Escape from GaNAs/GaAs Quantum Well Solar Cells. In Proceedings of the 2014 IEEE 40th Photovoltaic Specialist Conference (PVSC), Denver, CO, USA, 8–13 June 2014.
31. Tang, X.; Li, X.; Yue, C.; Wang, L.; Deng, Z.; Jia, H.; Wang, W.; Ji, A.; Jiang, Y.; Chen, H. Research on photo-generated carriers escape in PIN and NIN structures with quantum wells. *Appl. Phys. Express* **2020**, *13*, 071009. [CrossRef]
32. Busch, P.; Heinonen, T.; Lahti, P. Heisenberg's uncertainty principle. *Phys. Rep.* **2007**, *452*, 155–176. [CrossRef]

Disclaimer/Publisher's Note: The statements, opinions and data contained in all publications are solely those of the individual author(s) and contributor(s) and not of MDPI and/or the editor(s). MDPI and/or the editor(s) disclaim responsibility for any injury to people or property resulting from any ideas, methods, instructions or products referred to in the content.

Article

Structural, Surface, and Optical Properties of AlN Thin Films Grown on Different Substrates by PEALD

Sanjie Liu [1], Yangfeng Li [2], Jiayou Tao [1], Ruifan Tang [1,*] and Xinhe Zheng [3,*]

[1] Key Laboratory of Hunan Province on Information Photonics and Freespace Optical Communications, School of Physics and Electronic Science, Hunan Institute of Science and Technology, Yueyang 414006, China; liusanjie@hnist.edu.cn (S.L.); 12015013@hnist.edu.cn (J.T.)
[2] College of Semiconductors (College of Integrated Circuits), Hunan University, Changsha 410082, China; liyangfeng12@mails.ucas.ac.cn
[3] Beijing Key Laboratory for Magneto-Photoelectrical Composite and Interface Science, School of Mathematics and Physics, University of Science and Technology, Beijing 100083, China
* Correspondence: 12022055@hnist.edu.cn (R.T.); xinhezheng@ustb.edu.cn (X.Z.)

Abstract: Plasma-enhanced atomic layer deposition was employed to grow aluminum nitride (AlN) thin films on Si (100), Si (111), and c-plane sapphire substrates at 250 °C. Trimethylaluminum and $Ar/N_2/H_2$ plasma were utilized as Al and N precursors, respectively. The properties of AlN thin films grown on various substrates were comparatively analyzed. The investigation revealed that the as-grown AlN thin films exhibit a hexagonal wurtzite structure with preferred c-axis orientation and were polycrystalline, regardless of the substrates. The sharp AlN/substrate interfaces of the as-grown AlN are indicated by the clearly resolved Kiessig fringes measured through X-ray reflectivity. The surface morphology analysis indicated that the AlN grown on sapphire displays the largest crystal grain size and surface roughness value. Additionally, AlN/Si (100) shows the highest refractive index at a wavelength of 532 nm. Compared to AlN/sapphire, AlN/Si has a lower wavelength with an extinction coefficient of zero, indicating that AlN/Si has higher transmittance in the visible range. Overall, the study offers valuable insights into the properties of AlN thin films and their potential applications in optoelectronic devices, and provides a new technical idea for realizing high-quality AlN thin films with sharp AlN/substrate interfaces and smooth surfaces.

Keywords: aluminum nitride; thin films; different substrates; plasma-enhanced atomic layer deposition

1. Introduction

Aluminum nitride (AlN) is a highly promising wide-bandgap semiconductor because of its excellent thermal conductivity (single crystal: 285 W/m·K), along with its superior chemical stability and electrical insulation, which makes it a highly attractive material for high-electron-mobility transistors, light-emitting diodes, and laser diodes [1–4]. AlN is often utilized as a buffer layer for gallium nitride (GaN) in semiconductor device fabrication, reducing defects by providing a smooth and compatible interface between the substrate and the GaN layer [5–9]. Molecular beam epitaxy, sputtering, and metal–organic chemical vapor deposition are conventional methods used to grow high-quality AlN epilayers [10–13]. In general, these methods require elevated temperatures to achieve high-quality films and ensure adequate throughput. Unfortunately, high process temperatures yield additional stresses on the film stacks, which are not suitable for the fabrication of AlN films for sensory applications [14]. Therefore, alternative approaches are required to control the thickness of AlN films at relatively low temperatures for these specific applications.

Atomic layer deposition (ALD) is an attractive alternative low-temperature technique for preparing AlN thin films [15–20]. The deposition process relies on chemical surface reactions triggered by sequential precursor dosing. Precursors in the vapor phase chemisorb onto the surface via self-limiting reactions, wherein only the monolayer reacts [21–23]. This feature enables deposition on structures with high aspect ratios, facilitating the formation of thin films. Therefore, the film thickness can be accurately regulated by adjusting the number of reaction cycles. Plasma-enhanced ALD (PEALD) is a modified version of the conventional ALD technique used for depositing thin films of various materials with precise control over their thickness and composition. In PEALD, a plasma source is used to generate reactive species that can react with the precursor molecules and facilitate the deposition process. The plasma can also provide energy to the growing film, leading to lower deposition temperature, improved film quality, and enhanced step coverage [20].

In the literature, ALD and PEALD of AlN thin films have been extensively studied [24–31]. Polycrystalline AlN films were produced via both ALD and PEALD. The roughness of AlN films increased with higher deposition temperature and thicker film [31]. Tarala et al. obtained crystalline AlN using PEALD at temperatures less than 300 °C [32]. Shih et al. used ALD to deposit high-quality single-crystal hexagonal AlN, employing in-situ treatment. When these films were grown on GaN, the researchers observed increased mobility and sheet electron concentration [33]. Legallais et al. reported that adjusting ion energy through substrate biasing during the PEALD process significantly improved the quality of AlN [34]. Kim et al. deposited AlN on a GaN substrate using thermal ALD, and studied the interfacial properties of the ALD-AlN/GaN interface [35,36]. Bui et al. investigated the growth behavior and optical properties of ALD-AlN, revealing an increase in the refractive index with increasing film thickness and growth temperature [26,37]. Schiliro et al. obtained AlN with a predominant crystallographic orientation along the c-axis, with excellent alignment on GaN-sapphire substrates, resulting in high-density two-dimensional electron gas formation at the interface [38]. However, there have been relatively few studies conducted on how substrates impact the properties of AlN deposited via PEALD.

This study aimed to investigate and compare the characteristics of AlN films formed on different substrates, including Si (111) c-plane sapphire and Si (100), via PEALD at a temperature of 250 °C. The as-grown AlN films were polycrystalline, with a clear preference for the (002) orientation. Additionally, the interfaces between the films and substrates were extremely sharp. Surface morphology analysis revealed that the crystal grain size and surface roughness were highest for AlN films grown on sapphire, while the refractive index was highest for AlN/Si (100) at a wavelength of 532 nm. AlN/Si had a lower extinction coefficient than AlN/sapphire, indicating higher transmittance in the visible range.

2. Experimental

The deposition of AlN thin films was carried out using a PEALD process in an Angstrom-dep III reactor. The nitrogen and gallium precursors were trimethylaluminum (TMA) and a gas mixture containing Ar, N_2, and H_2 with a 1:3:6 ratio, respectively. The Si and sapphire substrates used in our experiments were wafers with a diameter of 2 inches. Prior to the growth experiments, we cleaned the sapphire substrates sequentially in an ultrasonic bath with isopropanol and acetone, followed by methanol and deionized water, while the Si was cleaned using the RCA standard cleaning process. The reactor was loaded with these substrates immediately, and the base pressure of the process was tuned to around 0.15 Torr. Subsequently, the substrate chuck was resistively heated to the growth temperature and given 20 min to reach thermal equilibrium. To enhance the crystalline quality and eliminate possible nucleation delays, the deposition cycles were preceded by a nitridation process step under 60 W rf-plasma power. $Ar/N_2/H_2$ (1:3:6) plasma was applied for 30 s, and the gas flow was set to 5 sccm. Then, deposition of AlN was carried out using the optimized PEALD-AlN parameters reported in our previous work [39], where the details are as follows: 15 s plasma/25 s N_2 purge /0.05 s TMA/25 s N_2 purge. During the experiment, the substrate temperature was maintained at 250 °C, and a total of 500 growth cycles were conducted.

X-ray photoelectron spectroscopy (XPS) was utilized to obtain information about the elemental composition, chemical state, and bonding of the AlN films. The instrument used was an ESCALAB 250Xi, which used an Al Kα X-ray source that was monochromatized.

Spectroscopic ellipsometry (SE) measurements were performed to determine the optical properties of the AlN films, covering the optical spectrum from 730 to 280 nm at an incidence angle of 70°. The spectra obtained from the measurements were analyzed to extract the thickness of the films. To assess the crystallinity of the GaN thin films, grazing incidence X-ray diffraction (GIXRD) was conducted. X-ray reflectometry (XRR) measurements were carried out to study the AlN/substrate interfacial property. Both GIXRD and XRR measurements are performed with a Smartlab system. The voltage and current were 40 kV and 30 mA, respectively. To enhance the intensity of X-rays diffracted from the thin film and avoid signals from the substrate, we conducted GIXRD measurements at a low grazing angle of 0.65°. Cross-sectional AlN was characterized using scanning electron microscopy (SEM, Zeiss supra55). The surface morphology of the samples was analyzed using the Micronano D-5A atomic force microscope (AFM) system. The scanning area was 1×1 um^2.

3. Results and Discussion

GIXRD analysis was conducted to determine the crystalline nature of the AlN thin films, and the obtained patterns (see Figure 1) exhibited diffraction peaks corresponding to the (100), (002), (101), and (110) planes of the hexagonal wurtzite phase, based on PDF#25-1133, indicating no mixing of other phases such as cubic. The analysis suggests that polycrystalline hexagonal AlN (h-AlN) films were obtained, displaying a dominant (002) orientation, regardless of the substrate used. Previous studies have suggested that ALD-deposited AlN films tend to crystallize in the (100) orientation [40,41], and sputtered AlN films that are more frequently reported to have c-plane orientation [42,43]. To obtain a preferred orientation of (002), the surface atoms must undergo kinetic energy-driven rearrangement to achieve a configuration with the minimum surface energy. When considering the steric hindrance of methyl groups, the low-density (100) plane could be even more beneficial because the mobility of atoms is diminished in low-temperature. Alevli et al. reported that AlN films exhibit (100) preferred orientation when grown at a low temperature of 185 °C, and the predominance of the (002) orientation becomes more prominent for AlN synthesized at elevated temperatures growth conditions. However, the AlN with a preferred orientation of (002) in the present work may be attributed to highly reactive radicals produced by Ar/N$_2$/H$_2$ plasma. The diffraction patterns of AlN samples exhibit similarities, but there is a difference in peak intensities: The (002) peak intensity in AlN/Si (111) was observed to be comparatively greater than that in AlN/Si (100) and AlN/sapphire. The location and width of the AlN (002) reflections in GIXRD are summarized in Table 1.

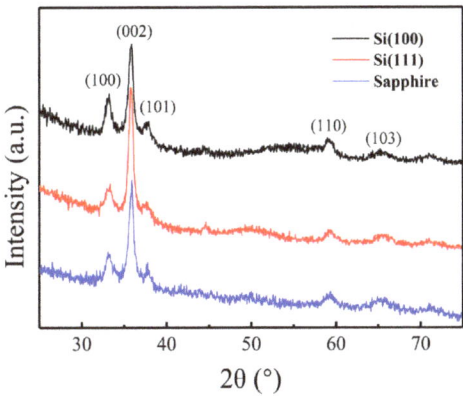

Figure 1. GIXRD patterns of AlN films grown on Si (100), Si (111), and sapphire substrates at 250 °C.

Table 1. GIXRD results of AlN on different substrates, and the grain size of the samples by the Scherrer equation.

Substrates	2θ (°)	FWHM (°)	c (Å)	a (Å)	Grain Size (Å)
Si (100)	35.941	0.575	4.992	3.105	146.86
Si (111)	35.755	0.554	5.018	3.100	152.35
c-Al$_2$O$_3$	35.899	0.530	4.999	3.122	159.31

The interplanar spacing, represented as "d", which corresponds to the crystal planes associated with the 2θ angles observed in XRD patterns, can be calculated by utilizing Bragg's equation for first-order diffraction [44]:

$$2d\sin\theta = n\lambda$$

where d can be expressed in terms of Miller indices (h, k, and l), "λ" denotes the X-ray wavelength (0.1540598 nm), and "θ" represents the angle of incidence.

The lattice constants "a" and "c" of hexagonal AlN can be determined using the following equation:

$$\frac{1}{d_{hkl}^2} = \frac{4}{3}\left(\frac{h^2+hk+k^2}{a^2}\right) + \frac{l^2}{c^2}$$

Using the (002) orientation, the lattice constants "c" for AlN/Si (100), AlN/Si (111), and AlN/sapphire were determined as 4.992 Å, 5.018 Å, and 4.999 Å, respectively. These values are slightly larger than the bulk AlN value of 4.981 Å [44], but consistent with previous reports on AlN grown by ALD [45]. The difference between the c-axis values of the as-grown AlN may be attributed to the mismatch in their crystal structures and thermal properties [46]. Using the (100) orientation, the lattice constants "a" for AlN/Si (100), AlN/Si (111), and AlN/sapphire were calculated as 3.105 Å, 3.100 Å, and 3.122 Å, respectively, which is comparable to the bulk AlN value of 3.111 Å [44]. The formation of defects during thin film growth could be one of the main reasons for the lattice constant variations. These defects can include dislocations, dislocation pile-ups, grain boundaries, and interface defects between the film and the substrate. These defects can have an impact on the crystal lattice structure of the thin film, resulting in changes in the lattice constants.

The disparities of lattice constants are related to film strain. In this case, the strains parallel to the c-axis [$\varepsilon^{\perp} = (c_{epi} - c_0)/c_0$] [47], for AlN grown on Si (100), Si (111), and sapphire, were determined to be 0.0022 GPa, 0.0074 Gpa, and 0.0036 GPa, respectively. This positive value indicates the presence of compressive strain, which means that the interplanar spacing along the c-axis is longer than that of unstrained AlN. The strains in the plane [$\varepsilon^{//} = (a_{epi} - a_0)/a_0$] were calculated as −0.0020 Gpa, −0.0030 Gpa and 0.0030 GPa, respectively. These negative values of $\varepsilon^{//}$ for AlN on Si suggest compressive strain in the in-plane directions, while the positive value for AlN on sapphire indicates tensile strain. This is because of the lattice mismatch of AlN between its substrates.

The average grain size of the sample was determined using the Scherrer equation [48], which relates the grain size (D) to the width of the peak at half of its maximum intensity (FWHM).

$$D = \frac{k\lambda}{B\cos\theta}$$

where D represents the average grain size (Å); k is the Scherrer constant, which is typically assumed to be 0.91, but may vary depending on the morphology of the crystal domains; λ is the X-ray wavelength, which depends on the type of X-ray used. In this study, the λ is 1.54058 Å (Cu Kα). B corresponds to the FWHM of the diffraction peak (radians), and θ represents the Bragg angle (radians). The average crystallite sizes for the (002) reflections of the AlN on Si (100), Si (111), and sapphire were calculated to be 146.86, 152.35, and 159.31 Å, respectively. The largest grain size of AlN/sapphire is indicated by the lowest FWHM of the (002) peak. Moreover, the average grain size of AlN/Si (111) was slightly larger than that of AlN/Si (100). Larger grain size usually relates to improved crystallinity [43].

The XRR results depicted in Figure 2 exhibit clear Kiessig fringes for all the AlN samples, suggesting the sharp interfaces between the AlN films and substrates. By fitting the oscillation period and amplitude, the film's thickness and density can be calculated. The experimental data were analyzed using Globfit software with two-layer models, i.e., AlN/Si and AlN/sapphire. The XRR data presented in Figure 2a exhibit excellent agreement between the measured and simulated results for the AlN/Si (111) sample. The measured film thicknesses were 84 nm, 72 nm, and 68 nm for AlN on Si (100), Si (111), and sapphire substrates, respectively. The variation in film thicknesses for AlN deposited on different substrates suggests different growth rates on each substrate. The initial growth of ALD is strongly influenced by the outermost surface of the substrate, as it determines the nucleation and initial bonding during the growth process [19]. The first step of plasma treatment, in the experiment, refers to the process of introducing nitrogen atoms into the surface of a material. In the case of sapphire and Si surfaces, it is generally easier to nitride the Si surface compared to the sapphire surface. The main reason for this difference is the chemical nature of the two materials [49]. Si has a higher affinity for nitrogen compared to sapphire, resulting in more adsorption sites (or reactive groups) on the Si surface, which facilitates the subsequent AlN deposition. The critical angle (θ_c), a material-specific property mainly dependent on film density, is located at the first minima of the second derivative of the XRR intensity [50]. In Figure 2b–d, the θ_c of the as-grown AlN thin films are presented in the insets. The extracted critical angles for AlN on Si (100) and Si (111) substrates were 0.230° and 0.238°, respectively, while for AlN on sapphire, it was 0.242°; these are a little higher than the reported values [51]. These critical angle values align with prior research on this topic. In addition, density values calculated from the simulations revealed that AlN/sapphire exhibits a density of 3.01 g/cm^3, which is lower than the bulk material's mass density of 3.25 g/cm^3, whereas those on Si (100) and Si (111) have densities of 2.81 and 2.86 g/cm^3, respectively. These results match well with the values previously reported for AlN films produced using various deposition techniques, such as PEALD and magnetron sputtering [52,53]. The result also suggests that the substrate can influence the density of the deposited AlN film. The higher density of AlN on sapphire compared to that on Si can be attributed to the lattice mismatch and interfacial interactions between N plasma and the respective substrates. Sapphire has a closer lattice matching with AlN compared to Si, allowing for better deposition and stronger interfacial bonding. This leads to a more compact and dense structure in the AlN film on sapphire. The higher density of AlN on Si (111) compared to Si (100) can be attributed to the better matching of hexagonal AlN with the Si (111) plane, which possesses six-fold symmetry, compared to the Si (100) plane with four-fold symmetry.

The cross-sectional SEM micrograph of AlN/Si (100) is shown in Figure 3, revealing an AlN film thickness of around 85 nm. This value is consistent with the one obtained from XRR measurements. In addition, the AlN on Si (100) exhibits a uniform morphology and a flat interface. The surface features of the AlN were investigated by AFM, and the images are presented in Figure 4. The roughness of film surface is commonly assessed using the root-mean-square (RMS) value of the surface roughness [54]. The RMS surface roughness values for AlN deposited on Si (111), Si (100), and sapphire substrates were determined to be 0.95, 1.02, and 1.36 nm, respectively. The obtained results are below the values previously reported for ALD-grown AlN [41]. By comparing the RMS surface roughness values and average crystallite sizes (Table 1) for the different substrates, we can observe that the AlN film on sapphire substrate has the highest RMS surface roughness value (1.36 nm), and also has the highest average grain size (159.31 Å). In contrast, AlN films on Si (111) substrates have the smallest RMS surface roughness value (0.95 nm) and the smallest average grain size (152.35 Å). This implies that there is a positive correlation between RMS surface roughness and average grain size, indicating that films with larger grain size usually yield higher surface roughness. The effect of substrate on surface roughness and grain growth is mainly due to the difference of lattice mismatch between AlN and substrate. The lattice mismatch between AlN and substrates is summarized in Table 2. In this case, AlN films on sapphire substrate show the largest RMS surface roughness and average grain size, possibly due to the better lattice match between substrate and AlN, which favors the growth of larger grains. However, AlN films on Si (111) and Si (100) substrates show lower surface roughness and smaller average grain size, which may be due to poor lattice matching.

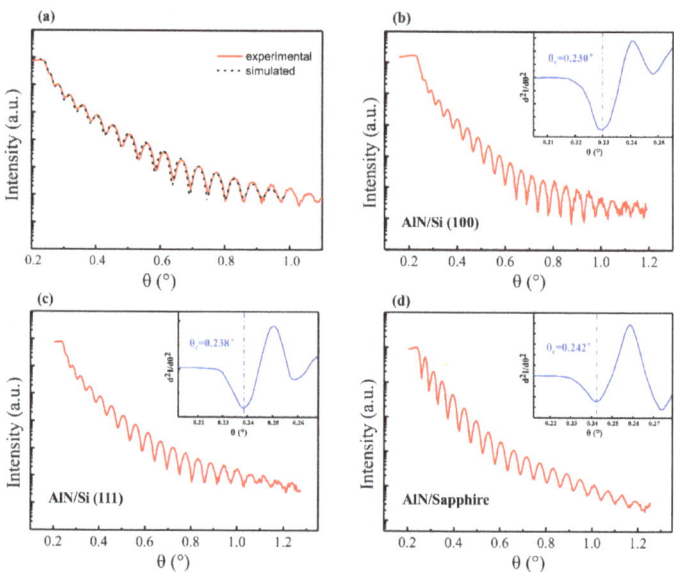

Figure 2. (a) Kiessig fringes of AlN on Si (111): solid and dotted lines illustrate experimental and fitted data, respectively. (b–d) X-ray reflectometry spectra for AlN thin films grown on Si (100), Si (111), and sapphire substrates, respectively.

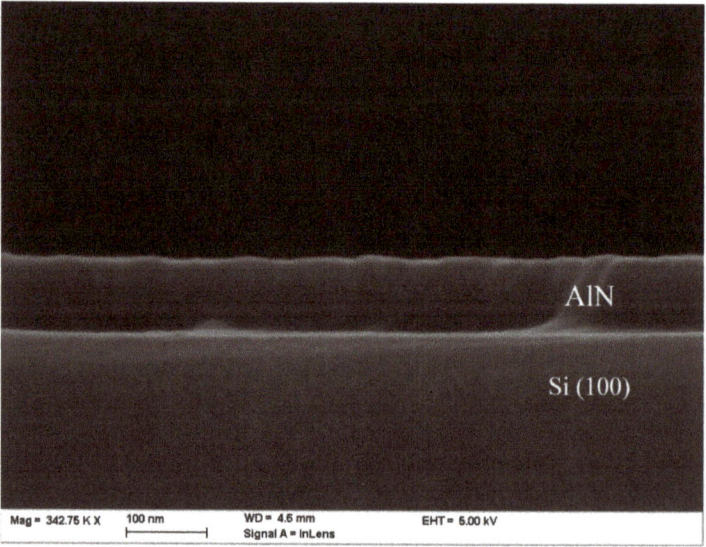

Figure 3. Cross-sectional SEM micrograph of the AlN film grown on Si (100).

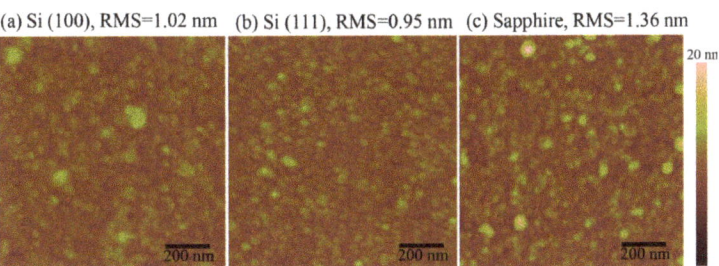

Figure 4. AFM image of AlN thin films grown on Si (100), Si (111), and sapphire substrates, respectively.

Table 2. The lattice parameters and lattice mismatch of AlN film and the substrates.

Materials	AlN	Sapphire	Si
Lattice constant (Å)	$a = 3.111$ $b = 4.981$	$a = 4.758$ $b = 12.99$	$a = 5.431$
Lattice Mismatch (with AlN)	—	+13.2%	−19.0%

Figure 5 presents a depth profile analysis of the AlN/Si (100). The results demonstrate that the atomic percentages of Al and N remain relatively constant throughout the film thickness in the etching direction. The Al and N contents in most AlN films are 47.2% and 52.0%, respectively, indicating that the membrane components are basically at the stoichiometric ratio. AlN contains a slightly higher proportion of nitrogen compared to aluminum. The higher nitrogen content in AlN is a result of the initial nitridation process, which involves the plasma-assisted breakdown of nitrogen gas [55]. The deposition process at a low temperature does not lead to the significant integration of carbon-containing ligands from the precursors, as indicated by the very low amount of carbon (0.7%) in the film. The content of oxygen in AlN is 1.1%, which is lower than the reported results [56], indicating negligible oxygen contamination. The main source of the high oxygen and carbon detected on the surface is atmospheric contamination.

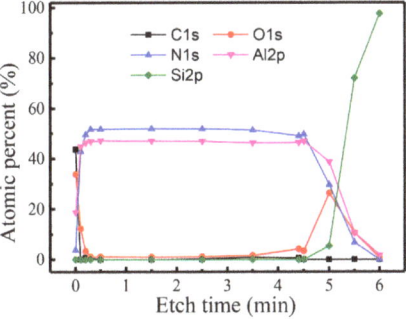

Figure 5. XPS depth profile of AlN thin films grown on Si (100).

The optical constants of AlN, such as refractive index (n) and extinction coefficient (k), were analyzed using SE. Figure 6 shows the dependence of n and k on the wavelength, ranging from 275 to 826 nm. The measurements reveal that at a wavelength of 532 nm, the n values of AlN films on Si (100), Si (111), and sapphire substrates are 1.967, 1.958, and 1.941, respectively. These values are below the bulk value ($n = 2.1$ at 533 nm) for single-crystal quality, but are consistent with the reported values of ALD-grown AlN [52]. In this case, AlN on Si has a higher n than that grown on sapphire. The wavelength of

$k = 0$, for AlN grown on both Si (100) and Si (111), is at 300 nm, with a rapid decrease between 275 and 300 nm. The zero of k value observed at higher wavelengths indicates that AlN/Si films remain transparent throughout the 300–826 nm range. Similarly, for AlN/sapphire, k decreases rapidly from 275 nm to 425 nm and becomes zero at 425 nm. Beyond this wavelength, k also remains at zero, indicating that AlN/sapphire films are transparent beyond a wavelength of 425 nm. These results suggest that AlN films grown on Si have a broader range of transparency wavelength than on sapphire substrates. The low absorption and high transparency are important factors for applications that require transparent materials in the ultraviolet range, such as photovoltaics, LEDs, and optical sensors. The transparency of the AlN films beyond these wavelengths also indicates that they may have potential for use as protective coatings or in optical devices that require high optical transparency in the visible range.

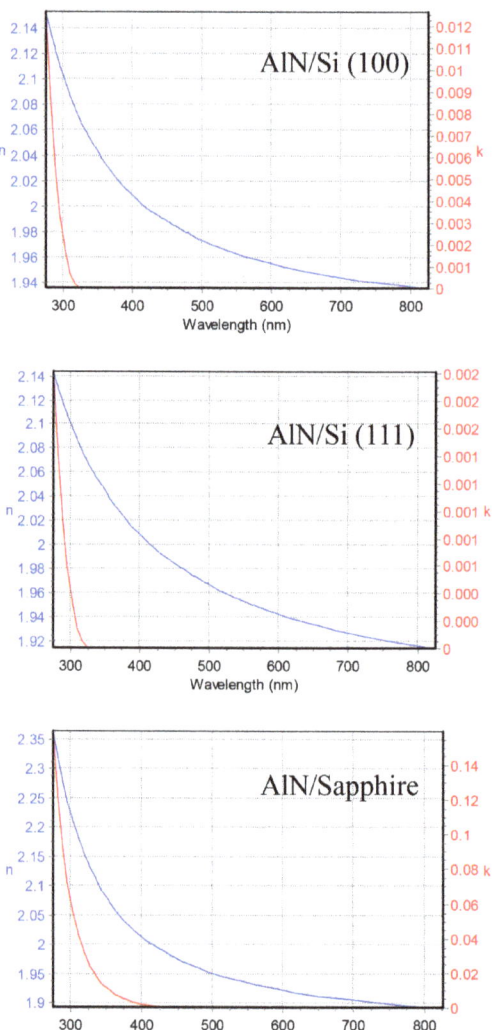

Figure 6. Refractive index and extinction coefficient of AlN films grown on different substrates as a function of wavelength.

4. Conclusions

In conclusion, the properties of PEALD-AlN deposited on Si (100), Si (111), and c-plane sapphire substrates were comparatively investigated. The results reveal that the as-grown AlN thin films possess a hexagonal wurtzite structure and are characterized as polycrystalline, regardless of the substrate. There is a certain relationship between the RMS surface roughness value and the average grain size, in which a larger RMS value is usually associated with a larger average grain size. The study also reveals that AlN/Si (100) has the highest n value at 532 nm, while AlN/sapphire exhibits a lower n value. Importantly, AlN films on both Si and sapphire substrates have a k value of zero within the visible range, indicating an optical transparency in the visible range. These results suggest that AlN thin films have significant potential for optoelectronic applications where optical transparency is a critical factor. This work of obtaining AlN films with smooth surfaces and sharp AlN/substrate interfaces is essential for their use as buffer layers in heteroepitaxy GaN.

Author Contributions: S.L. conceived and executed the project. Y.L. contributed XRR measurements and reviewed the manuscript. X.Z., J.T. and R.T. provided funding support for the project. S.L. performed the thin film deposition and XRD, SEM, and SE measurements; analyzed the data; and wrote the manuscript. All authors have read and agreed to the published version of the manuscript.

Funding: This work was supported by the Hunan Provincial Natural Science Foundation of China (Grant No. 2022JJ40163, Grant No. 2021JJ30298, Grant No. 2023JJ50283), the Education Department of Hunan Province (Grant No. 21B0598, Grant No. 22B0020), the Natural Science Foundation of China (Grant No. 52002021), Fundamental Research Funds for the Central Universities (Grant No. FRF-TP-20-016A2, Grant No. FRF-BR-20-02A), and Natural Science Foundation of Changsha (Grant No. kq2208213).

Data Availability Statement: The data that support the findings of this study are available from the corresponding author upon reasonable request.

Acknowledgments: We thank Gang Zhao from the School of Physics and Electronics, Hunan Normal University, for the XPS measurements.

Conflicts of Interest: The authors declare no conflict of interest.

References

1. Taniyasu, Y.; Kasu, M.; Makimoto, T. An aluminium nitride light-emitting diode with a wavelength of 210 nanometres. *Nature* **2006**, *441*, 325–328. [CrossRef]
2. Lu, J.; Chen, J.-T.; Dahlqvist, M.; Kabouche, R.; Medjdoub, F.; Rosen, J.; Kordina, O.; Hultman, L. Transmorphic epitaxial growth of AlN nucleation layers on SiC substrates for high-breakdown thin GaN transistors. *Appl. Phys. Lett.* **2019**, *115*, 221601. [CrossRef]
3. Zhang, D.; Cheng, X.; Shen, L.; Zheng, L.; Gu, Z.; Zhou, W.; Liu, X.; Yu, Y. Influence of Poly-AlN Passivation on the Performance Improvement of 3-MeV Proton-Irradiated AlGaN/GaN MIS-HEMTs. *IEEE Trans. Nucl. Sci.* **2019**, *66*, 2215–2219. [CrossRef]
4. Ambacher, O. Growth and applications of Group III-nitrides. *J. Phys. D Appl. Phys.* **1998**, *31*, 2653. [CrossRef]
5. Amano, H.; Sawaki, N.; Akasaki, I.; Toyoda, Y. Metalorganic vapor phase epitaxial growth of a high quality GaN film using an AlN buffer layer. *Appl. Phys. Lett.* **1986**, *48*, 353–355. [CrossRef]
6. Liu, S.; Yang, S.; Tang, Z.; Jiang, Q.; Liu, C.; Wang, M.; Shen, B.; Chen, K.J. Interface/border trap characterization of Al_2O_3/AlN/GaN metal-oxide-semiconductor structures with an AlN interfacial layer. *Appl. Phys. Lett.* **2015**, *106*, 051605. [CrossRef]
7. Pan, L.; Dong, X.; Li, Z.; Luo, W.; Ni, J. Influence of the AlN nucleation layer on the properties of AlGaN/GaN heterostructure on Si (1 1 1) substrates. *Appl. Surf. Sci.* **2018**, *447*, 512–517. [CrossRef]
8. Liudi Mulyo, A.; Rajpalke, M.K.; Vullum, P.E.; Weman, H.; Kishino, K.; Fimland, B.-O. The influence of AlN buffer layer on the growth of self-assembled GaN nanocolumns on graphene. *Sci. Rep.* **2020**, *10*, 853. [CrossRef]
9. Zhan, X.; Liu, J.; Sun, X.; Huang, Y.; Gao, H.; Zhou, Y.; Li, Q.; Sun, Q.; Yang, H. Crack-free 2.2 µm-thick GaN grown on Si with a single-layer AlN buffer for RF device applications. *J. Phys. D Appl. Phys.* **2023**, *56*, 015104. [CrossRef]
10. Yeadon, M.; Marshall, M.T.; Hamdani, F.; Pekin, S.; Morkoç, H.; Gibson, J.M. In situ transmission electron microscopy of AlN growth by nitridation of (0001) α-Al_2O_3. *J. Appl. Phys.* **1998**, *83*, 2847–2850. [CrossRef]
11. Zscherp, M.F.; Mengel, N.; Hofmann, D.M.; Lider, V.; Ojaghi Dogahe, B.; Becker, C.; Beyer, A.; Volz, K.; Schörmann, J.; Chatterjee, S. AlN Buffer Enhances the Layer Quality of MBE-Grown Cubic GaN on 3C-SiC. *Cryst. Growth Des.* **2022**, *22*, 6786–6791. [CrossRef]

12. Lutsenko, E.V.; Rzheutski, M.V.; Vainilovich, A.G.; Svitsiankou, I.E.; Shulenkova, V.A.; Muravitskaya, E.V.; Alexeev, A.N.; Petrov, S.I.; Yablonskii, G.P. MBE AlGaN/GaN HEMT Heterostructures with Optimized AlN Buffer on Al_2O_3. *Semiconductors* **2018**, *52*, 2107–2110. [CrossRef]
13. Zamir, S.; Meyler, B.; Zolotoyabko, E.; Salzman, J. The effect of AlN buffer layer on GaN grown on (111)-oriented Si substrates by MOCVD. *J. Cryst. Growth* **2000**, *218*, 181–190. [CrossRef]
14. Yarar, E.; Hrkac, V.; Zamponi, C.; Piorra, A.; Kienle, L.; Quandt, E. Low temperature aluminum nitride thin films for sensory applications. *AIP Adv.* **2016**, *6*, 075115. [CrossRef]
15. George, S.M. Atomic Layer Deposition: An Overview. *Chem. Rev.* **2010**, *110*, 111–131. [CrossRef] [PubMed]
16. Cremers, V.; Puurunen, R.L.; Dendooven, J. Conformality in atomic layer deposition: Current status overview of analysis and modelling. *Appl. Phys. Rev.* **2019**, *6*, 021302. [CrossRef]
17. Vervuurt, R.H.J.; Karasulu, B.; Verheijen, M.A.; Kessels, W.M.M.; Bol, A.A. Uniform Atomic Layer Deposition of Al_2O_3 on Graphene by Reversible Hydrogen Plasma Functionalization. *Chem. Mater.* **2017**, *29*, 2090–2100. [CrossRef]
18. Puurunen, R.L. Correlation between the growth-per-cycle and the surface hydroxyl group concentration in the atomic layer deposition of aluminum oxide from trimethylaluminum and water. *Appl. Surf. Sci.* **2005**, *245*, 6–10. [CrossRef]
19. Miikkulainen, V.; Leskelä, M.; Ritala, M.; Puurunen, R. ChemInform Abstract: Crystallinity of Inorganic Films Grown by Atomic Layer Deposition: Overview and General Trends. *J. Appl. Phys.* **2013**, *113*, 2. [CrossRef]
20. Profijt, H.B.; Potts, S.E.; van de Sanden, M.C.M.; Kessels, W.M.M. Plasma-Assisted Atomic Layer Deposition: Basics, Opportunities, and Challenges. *J. Vac. Sci. Technol. A* **2011**, *29*, 050801. [CrossRef]
21. Gakis, G.P.; Vahlas, C.; Vergnes, H.; Dourdain, S.; Tison, Y.; Martinez, H.; Bour, J.; Ruch, D.; Boudouvis, A.G.; Caussat, B.; et al. Investigation of the initial deposition steps and the interfacial layer of Atomic Layer Deposited (ALD) Al_2O_3 on Si. *Appl. Surf. Sci.* **2019**, *492*, 245–254. [CrossRef]
22. Lu, J.; Elam, J.W.; Stair, P.C. Atomic layer deposition—Sequential self-limiting surface reactions for advanced catalyst "bottom-up" synthesis. *Surf. Sci. Rep.* **2016**, *71*, 410–472. [CrossRef]
23. Zaera, F. Mechanisms of surface reactions in thin solid film chemical deposition processes. *Coord. Chem. Rev.* **2013**, *257*, 3177–3191. [CrossRef]
24. Lee, Y.J.; Kang, S.-W. Growth of aluminum nitride thin films prepared by plasma-enhanced atomic layer deposition. *Thin Solid Film.* **2004**, *446*, 227–231. [CrossRef]
25. Ozgit, C.; Donmez, I.; Alevli, M.; Biyikli, N. Self-limiting low-temperature growth of crystalline AlN thin films by plasma-enhanced atomic layer deposition. *Thin Solid Film.* **2012**, *520*, 2750–2755. [CrossRef]
26. Van Bui, H.; Wiggers, F.B.; Gupta, A.; Nguyen, M.D.; Aarnink, A.A.I.; de Jong, M.P.; Kovalgin, A.Y. Initial growth, refractive index, and crystallinity of thermal and plasma-enhanced atomic layer deposition AlN films. *J. Vac. Sci. Technol. A* **2014**, *33*, 01A111. [CrossRef]
27. Banerjee, S.; Aarnink, A.A.I.; van de Kruijs, R.; Kovalgin, A.Y.; Schmitz, J. PEALD AlN: Controlling growth and film crystallinity. *Phys. Status Solidi C* **2015**, *12*, 1036–1042. [CrossRef]
28. Bosund, M.; Sajavaara, T.; Laitinen, M.; Huhtio, T.; Putkonen, M.; Airaksinen, V.-M.; Lipsanen, H. Properties of AlN grown by plasma enhanced atomic layer deposition. *Appl. Surf. Sci.* **2011**, *257*, 7827–7830. [CrossRef]
29. Kim, Y.; Kim, M.S.; Yun, H.J.; Ryu, S.Y.; Choi, B.J. Effect of growth temperature on AlN thin films fabricated by atomic layer deposition. *Ceram. Int.* **2018**, *44*, 17447–17452. [CrossRef]
30. Nguyen, T.; Adjeroud, N.; Glinsek, S.; Fleming, Y.; Guillot, J.; Grysan, P.; Polesel-Maris, J. A film-texture driven piezoelectricity of AlN thin films grown at low temperatures by plasma-enhanced atomic layer deposition. *APL Mater.* **2020**, *8*, 071101. [CrossRef]
31. Liu, X.; Ramanathan, S.; Lee, E.; Seidel, T.E. Atomic Layer Deposition of Aluminum Nitride Thin films from Trimethyl Aluminum (TMA) and Ammonia. *MRS Online Proc. Libr.* **2003**, *811*, 158–163.
32. Tarala, V.; Altakhov, A.; Martens, V.; Lisitsyn, S. Growing aluminum nitride films by Plasma-Enhanced Atomic Layer Deposition at low temperatures. *J. Phys. Conf. Ser.* **2015**, *652*, 012034. [CrossRef]
33. Shih, H.-Y.; Lee, W.-H.; Kao, W.-C.; Chuang, Y.-C.; Lin, R.-M.; Lin, H.-C.; Shiojiri, M.; Chen, M.-J. Low-temperature atomic layer epitaxy of AlN ultrathin films by layer-by-layer, in-situ atomic layer annealing. *Sci. Rep.* **2017**, *7*, 39717. [CrossRef]
34. Legallais, M.; Mehdi, H.; David, S.; Bassani, F.; Labau, S.; Pelissier, B.; Baron, T.; Martinez, E.; Ghibaudo, G.; Salem, B. Improvement of AlN Film Quality Using Plasma Enhanced Atomic Layer Deposition with Substrate Biasing. *ACS Appl. Mater. Interfaces* **2020**, *12*, 39870–39880. [CrossRef]
35. Kim, H.; Kim, N.D.; An, S.C.; Yoon, H.J.; Choi, B.J. Improved interfacial properties of thermal atomic layer deposited AlN on GaN. *Vacuum* **2019**, *159*, 379–381. [CrossRef]
36. Kim, H.; Yun, H.J.; Choi, S.; Choi, B.J. Comparison of electrical and interfacial characteristics between atomic-layer-deposited AlN and AlGaN on a GaN substrate. *Appl. Phys. A* **2020**, *126*, 449. [CrossRef]
37. Van Bui, H.; Nguyen, M.D.; Wiggers, F.B.; Aarnink, A.A.I.; de Jong, M.P.; Kovalgin, A.Y. Self-Limiting Growth and Thickness- and Temperature- Dependence of Optical Constants of ALD AlN Thin Films. *ECS J. Solid State Sci. Technol.* **2014**, *3*, P101. [CrossRef]
38. Schilirò, E.; Giannazzo, F.; Bongiorno, C.; Di Franco, S.; Greco, G.; Roccaforte, F.; Prystawko, P.; Kruszewski, P.; Leszczyński, M.; Krysko, M.; et al. Structural and electrical properties of AlN thin films on GaN substrates grown by plasma enhanced-Atomic Layer Deposition. *Mater. Sci. Semicond. Process.* **2019**, *97*, 35–39. [CrossRef]

39. Liu, S.; Peng, M.; Hou, C.; He, Y.; Li, M.; Zheng, X. PEALD-Grown Crystalline AlN Films on Si (100) with Sharp Interface and Good Uniformity. *Nanoscale Res. Lett.* **2017**, *12*, 279. [CrossRef]
40. Lei, W.; Chen, Q. Crystal AlN deposited at low temperature by magnetic field enhanced plasma assisted atomic layer deposition. *J. Vac. Sci. Technol. A* **2012**, *31*, 01A114. [CrossRef]
41. Alevli, M.; Ozgit, C.; Donmez, I.; Biyikli, N. The influence of N2/H2 and ammonia N source materials on optical and structural properties of AlN films grown by plasma enhanced atomic layer deposition. *J. Cryst. Growth* **2011**, *335*, 51–57. [CrossRef]
42. Iriarte, G.F.; Reyes, D.F.; González, D.; Rodriguez, J.G.; García, R.; Calle, F. Influence of substrate crystallography on the room temperature synthesis of AlN thin films by reactive sputtering. *Appl. Surf. Sci.* **2011**, *257*, 9306–9313. [CrossRef]
43. García-Méndez, M.; Morales-Rodríguez, S.; Shaji, S.; Krishnan, B.; Bartolo-Pérez, P. Structural properties of AlN films with oxygen content deposited by reactive magnetron sputtering: XRD and XPS characterization. *Surf. Rev. Lett.* **2011**, *18*, 23–31. [CrossRef]
44. Moram, M.A.; Vickers, M.E. X-ray diffraction of III-nitrides. *Rep. Prog. Phys.* **2009**, *72*, 036502. [CrossRef]
45. Österlund, E.; Seppänen, H.; Bespalova, K.; Miikkulainen, V.; Paulasto-Kröckel, M. Atomic layer deposition of AlN using atomic layer annealing—Towards high-quality AlN on vertical sidewalls. *J. Vac. Sci. Technol. A* **2021**, *39*, 032403. [CrossRef]
46. Sun, C.J.; Kung, P.; Saxler, A.; Ohsato, H.; Haritos, K.; Razeghi, M. A crystallographic model of (00·1) aluminum nitride epitaxial thin film growth on (00·1) sapphire substrate. *J. Appl. Phys.* **1994**, *75*, 3964–3967. [CrossRef]
47. Zhou, S.Q.; Vantomme, A.; Zhang, B.S.; Yang, H.; Wu, M.F. Comparison of the properties of GaN grown on complex Si-based structures. *Appl. Phys. Lett.* **2005**, *86*, 081912. [CrossRef]
48. Scherrer, P. Bestimmung der Grösse und der inneren Struktur von Kolloidteilchen mittels Röntgenstrahlen. *Nachr. Von Der Ges. Der Wiss. Zu Göttingen Math. Phys. Kl.* **1918**, *1918*, 98–100.
49. Dovidenko, K.; Oktyabrsky, S.; Narayan, J.; Razeghi, M. Aluminum nitride films on different orientations of sapphire and silicon. *J. Appl. Phys.* **1996**, *79*, 2439–2445. [CrossRef]
50. Chason, E.; Mayer, T.M. Thin film and surface characterization by specular X-ray reflectivity. *Crit. Rev. Solid State Mater. Sci.* **1997**, *22*, 1–67. [CrossRef]
51. Motamedi, P.; Cadien, K. Structural and optical characterization of low-temperature ALD crystalline AlN. *J. Cryst. Growth* **2015**, *421*, 45–52. [CrossRef]
52. Alevli, M.; Ozgit, C.; Donmez, I.; Biyikli, N. Structural properties of AlN films deposited by plasma-enhanced atomic layer deposition at different growth temperatures. *Phys. Status Solidi A* **2012**, *209*, 266–271. [CrossRef]
53. Venkataraj, S.; Severin, D.; Drese, R.; Koerfer, F.; Wuttig, M. Structural, optical and mechanical properties of aluminium nitride films prepared by reactive DC magnetron sputtering. *Thin Solid Film.* **2006**, *502*, 235–239. [CrossRef]
54. Martin, Y.; Williams, C.C.; Wickramasinghe, H.K. Atomic force microscope–force mapping and profiling on a sub 100-Å scale. *J. Appl. Phys.* **1987**, *61*, 4723–4729. [CrossRef]
55. Liu, S.; Zhao, G.; He, Y.; Wei, H.; Li, Y.; Qiu, P.; Song, Y.; An, Y.; Wang, X.; Wang, X.; et al. Interfacial Tailoring for the Suppression of Impurities in GaN by In Situ Plasma Pretreatment via Atomic Layer Deposition. *ACS Appl. Mater. Interfaces* **2019**, *11*, 35382–35388. [CrossRef]
56. Motamedi, P.; Cadien, K. XPS analysis of AlN thin films deposited by plasma enhanced atomic layer deposition. *Appl. Surf. Sci.* **2014**, *315*, 104–109. [CrossRef]

Disclaimer/Publisher's Note: The statements, opinions and data contained in all publications are solely those of the individual author(s) and contributor(s) and not of MDPI and/or the editor(s). MDPI and/or the editor(s) disclaim responsibility for any injury to people or property resulting from any ideas, methods, instructions or products referred to in the content.

Article

Crack-Free High-Composition (>35%) Thick-Barrier (>30 nm) AlGaN/AlN/GaN High-Electron-Mobility Transistor on Sapphire with Low Sheet Resistance (<250 Ω/□)

Swarnav Mukhopadhyay *, Cheng Liu, Jiahao Chen, Md Tahmidul Alam, Surjava Sanyal, Ruixin Bai, Guangying Wang, Chirag Gupta and Shubhra S. Pasayat

Electrical & Computer Engineering, University of Wisconsin-Madison, Madison, WI 53706, USA; cliu634@wisc.edu (C.L.); jchen967@wisc.edu (J.C.); malam9@wisc.edu (M.T.A.); ssanyal2@wisc.edu (S.S.); rbai33@wisc.edu (R.B.); gwang265@wisc.edu (G.W.); cgupta9@wisc.edu (C.G.); shubhra@ece.wisc.edu (S.S.P.)
* Correspondence: swarnav.mukhopadhyay@wisc.edu

Abstract: In this article, a high-composition (>35%) thick-barrier (>30 nm) AlGaN/AlN/GaN high-electron-mobility transistor (HEMT) structure grown on a sapphire substrate with ultra-low sheet resistivity (<250 Ω/□) is reported. The optimization of growth conditions, such as reduced deposition rate, and the thickness optimization of different epitaxial layers allowed us to deposit a crack-free high-composition and thick AlGaN barrier layer HEMT structure. A significantly high two-dimensional electron gas (2DEG) density of 1.46×10^{13} cm^{-2} with a room-temperature mobility of 1710 cm^2/V·s was obtained via Hall measurement using the Van der Pauw method. These state-of-the-art results show great potential for high-power Ga-polar HEMT design on sapphire substrates.

Keywords: high-electron-mobility transistors; ultra-low sheet resistance; 2DEG; MOCVD; electron mobility; crack-free AlGaN; high-composition AlGaN barrier

Citation: Mukhopadhyay, S.; Liu, C.; Chen, J.; Tahmidul Alam, M.; Sanyal, S.; Bai, R.; Wang, G.; Gupta, C.; Pasayat, S.S. Crack-Free High-Composition (>35%) Thick-Barrier (>30 nm) AlGaN/AlN/GaN High-Electron-Mobility Transistor on Sapphire with Low Sheet Resistance (<250 Ω/□). *Crystals* **2023**, *13*, 1456. https://doi.org/10.3390/cryst13101456

Academic Editors: Yangfeng Li, Zeyu Liu, Mingzeng Peng, Yang Wang, Yang Jiang and Yuanpeng Wu

Received: 31 August 2023
Revised: 22 September 2023
Accepted: 26 September 2023
Published: 30 September 2023

Copyright: © 2023 by the authors. Licensee MDPI, Basel, Switzerland. This article is an open access article distributed under the terms and conditions of the Creative Commons Attribution (CC BY) license (https://creativecommons.org/licenses/by/4.0/).

1. Introduction

The deposition of III-Nitride materials has advanced in the past two decades. The accelerated development of high-electron-mobility transistors (HEMTs) can be attributed to the in-depth investigations conducted on III-Nitride materials. The ability of AlGaN/GaN HEMTs to operate at high voltage while maintaining a low on-resistance (R_{ON}) makes them attractive for power electronics applications. The growing interest as well as commercialization of AlGaN/GaN HEMTs in the power electronics industry requires continuous development to enhance their performance beyond the current state of the art. One of the most challenging aspects of advancing AlGaN/GaN HEMT technology is the deposition of a crack-free high-composition and thick AlGaN barrier layer. This combination is needed to reduce gate leakage and to enable an increased breakdown voltage while maintaining a low sheet resistance [1,2]. This is especially significant for high-voltage operation, where the high electric field at the drain side gate edge often leads to an excessive gate leakage, causing soft breakdown. In addition to addressing gate leakage, the heightened critical electric field of the high-composition AlGaN barrier layer contributes to enhancing the breakdown voltage of the HEMT. The limitation of depositing a high-composition thick AlGaN barrier layer arises from the critical layer thickness of the AlGaN barrier layer on thick GaN buffer layers due to a 3.5% lattice mismatch between AlN and GaN [3]. Upon exceeding the critical layer thickness, the AlGaN layer tends to crack due to the high tensile strain in the film when deposited on the GaN base layers. Concurrently, a higher AlGaN barrier layer thickness results in higher polarization-induced charges in the channel, which are required to obtain low sheet resistance. Another crucial challenge is that the high-composition thick AlGaN barrier layer shifts the 2DEG wavefunction towards the interface, causing increased alloy scattering and reduced electron mobility [4].

Moreover, with increased 2DEG density, carrier–carrier scattering increases significantly in the channel, thus limiting high electron mobility [5]. Therefore, it is difficult to grow a sufficiently thick high-composition AlGaN barrier layer for an HEMT structure with high 2DEG density and simultaneously achieve the high mobility that is necessary for power electronic devices. Additionally, most of the high-performance AlGaN/GaN HEMTs with room-temperature sheet resistivity near 250 Ω/□ have been demonstrated with SiC substrates, which are cost-limiting for power electronics applications [6–8]. There have been very few reports of AlGaN/GaN HEMTs with sheet resistivity of near or less than 250 Ω/□ [6,8] with a single AlGaN barrier, with most of them on SiC substrates. It is challenging to achieve low sheet resistivity in AlGaN/GaN HEMT structures deposited on sapphire substrates due to the high density of defects and dislocations generated due to lattice mismatch (16%) between GaN and sapphire [9,10]. Also, it is well known that due to high optical phonon scattering and scattering due to polarization-induced charges at the interface, it is very difficult to achieve sheet resistivity below 250 Ω/□ [11]. The lowest recorded sheet resistivity in an AlGaN/GaN HEMT on SiC so far is 211 Ω/□, achieved by Yamada et al. [6] using a high-composition $Al_{0.68}Ga_{0.32}N$ (layer thickness 6 nm) barrier layer. Moreover, maintaining a sharp and smooth interface is crucial for obtaining high electron mobility, which is difficult for high-composition AlGaN barrier layers [4,6]. The interface quality will degrade as the thickness of the high-composition AlGaN barrier layer is increased due to strain, which represents a challenge [4]. Although some of the literature has shown >30% AlGaN barrier GaN HEMTs with >20 nm barrier thickness having sheet resistance >250 Ω/□ [12–15], however achieving >30 nm barrier thickness with >35% Al composition with <250 Ω/□ is non-trivial. As a result, solving these issues would have great implications in terms of improving the performance of high-voltage HEMT devices.

Recently, novel HEMT designs such as multi-channel GaN HEMTs are showing great potential for achieving high breakdown voltagea (>1000 V) with a low sheet resistance (<150 Ω/□) [16,17]. Also, novel gate designs such as p-GaN-gated HEMTs have shown excellent performance in reducing gate leakage currents to less than 1 μA/mm and increasing breakdown voltages to >1000 V [18–20].

Furthermore, different design techniques such as back-barrier design using an AlGaN back barrier and a C-doped buffer layer are reported to enhance the breakdown voltage beyond 1 KV in AlGaN/GaN HEMTs [21–23]. However, C-doping can cause degradation in material quality and cause degradation of 2DEG mobility due to ionized impurity scattering [21], whereas AlGaN back-barrier design can introduce more strain into the HEMT structure due to increased lattice mismatch and deteriorate the material quality by introducing more dislocation density [24]. So, in this work, the authors tried to show a pathway towards achieving high breakdown voltage with a lower sheet resistance in a simple way by increasing the Al composition and thickness of the barrier layer.

In this manuscript, the authors present the AlGaN/AlN/GaN HEMT structure deposited on a c-plane sapphire substrate with 36% Al composition and 31 nm barrier thickness, with sheet resistance as low as 249 Ω/□ at room temperature and with a very high mobility of 7830 $cm^2/V·s$ at a cryogenic temperature (77 K), which indicates a sharp and smooth interface. An optimized deposition process is developed to address all the issues that have been discussed above, and the results show that further development of Ga-polar HEMTs is needed and that it is possible to overcome the intrinsic limits of the sheet resistance of AlGaN/GaN HEMTs with improved deposition techniques.

2. Experimental Methods

An AlGaN/AlN/GaN HEMT was deposited using MOCVD on standard Fe-doped GaN on c-plane sapphire templates using tri-methyl gallium (TMGa), tri-ethyl gallium (TEGa), and tri-methyl aluminum (TMAl) as metal–organic precursors along with ammonia (NH_3) as the group-V precursor. H_2 was used as a carrier gas. A thick UID-GaN layer (thickness t_1) was initially deposited on a standard Fe-doped semi-insulating GaN on sapphire templates using 90 μmol/min of TMGa and 283 mmol/min of NH_3. Next, a

40 nm GaN channel was deposited using TEGa followed by a thin AlN layer (thickness t_2) together with a thick $Al_{0.36}Ga_{0.64}N$ layer (thickness t_3) with a V/III ratio of 3800 at 1210 °C throughout these layers. The TMAl flow was 4.6 µmol/min and the TEGa flow was 15 µmol/min.

For obtaining a thick and high-composition AlGaN/AlN/GaN HEMT with a low sheet resistance (<250 Ω/□), a series of experiments were performed to optimize the deposition conditions and understand the effect of different deposition parameters on 2DEG mobility. Three different deposition temperatures (1100 °C, 1210 °C, and 1270 °C) were examined initially to obtain the most optimized deposition condition in the GaN channel. Multiple structural parameters were varied—the thickness t_1 of the intermediate UID-GaN layer (to understand the effect of the distance of the channel from the deposition surface of the Fe-doped semi-insulating GaN template), the thickness t_2 of the AlN layer (to check the effect of alloy scattering), and finally the thickness t_3 of the AlGaN layer (to study its effect on the channel resistance). The final epitaxial structure design included the most optimized deposition conditions obtained by varying all of the above parameters. The epitaxial layer structure is shown in Figure 1. The Al composition in the barrier layer was kept constant (36%) throughout the different epitaxial layers to obtain enhanced polarization-induced 2DEG charge density. The composition of the AlGaN barrier layer was kept at 36% to minimize the sheet resistance while depositing the thick barrier layer to avoid the formation of cracks. The strain in AlGaN layers deposited on relaxed GaN continued increasing with the increase in Al composition and thickness, which increases the probability of crack formation [25]; so, in this work, the composition was not increased beyond 36%. Also, it has been reported that an Al composition greater than or equal to 38% reduces the low-temperature (77 K) 2DEG mobility, which signifies increased interface roughness scattering in the channel [26,27]. Deposition parameters are listed in Table 1. The UID GaN thickness (t_1) was chosen to reduce the effect of unintentional Fe doping in the UID-GaN layer from the Fe-doped S.I. GaN template, known as the memory effect of Fe in GaN [28,29]. Thus, we have chosen two different thicknesses of UID GaN layers, 200 nm and 1000 nm, for observing the effect of unintentional Fe doping from the S.I. template. The high-composition AlGaN barrier and the AlN interlayer thickness were designed to increase 2DEG mobility and minimize the sheet resistance of the channel. The AlN thickness was chosen as 0.7 nm and 1.2 nm, as a thickness of AlN below 0.7 nm does not reduce the alloy disorder scattering significantly. Moreover, an AlN thickness beyond 1.2 nm will cause micro-cracks in the barrier layer due to increased strain. The AlGaN thicknesses were chosen to be 21 nm and 31 nm for determining the increment in charge density and the effect on 2DEG mobility. An intermediate thickness of 25 nm of AlGaN was also deposited, but due to a drift in the deposition condition the Al composition in the AlGaN layer changed, thus the 2DEG charge density and the mobility values could not be compared to the other samples.

GaN deposited using TEGa at temperatures ≥1000 °C is known to show a 20–40% lower FWHM in the XRD omega-rocking curve and a greater than 30% improvement in electron mobility compared to TMGa [30–32]. Also, the deposition rate can be precisely controlled over less than 1 Å/s with the TEGa precursor due to its 10–50 times lower vapor pressure with reference to TMGa [33]. Yamada et al. have shown that controlled deposition of a high-composition (>35%) AlGaN barrier and GaN channel using TEGa can enhance the 2DEG mobility in HEMTs up to 2000 $cm^2/V \cdot s$ [6], even though the deposition rate of GaN using TEGa is very low compared to TMGa. This can lead to a potential 10-fold increase in the deposition time of GaN HEMT. So, we used the advantages of both TMGa and TEGa in this work, by depositing a thick GaN buffer using TMGa and a thin GaN channel (40 nm) along with an AlGaN barrier with TEGa for superior crystalline quality [32].

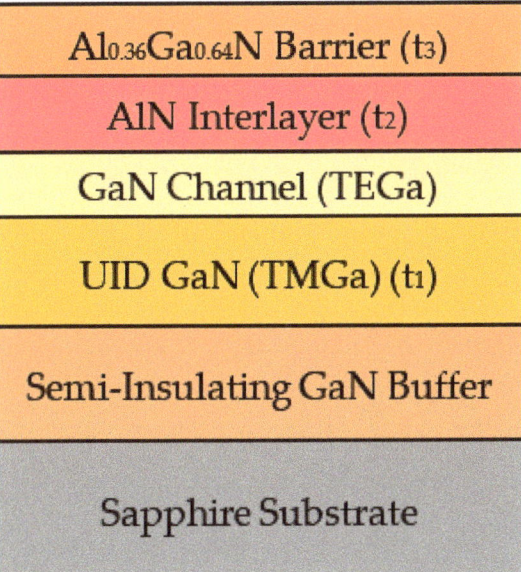

Figure 1. Epitaxial layer of AlGaN/AlN/GaN HEMT structure.

Table 1. Deposition parameters of different AlGaN/AlN/GaN HEMT samples.

Sample No.	UID-GaN Thickness (nm) (t_1)	AlN Thickness (nm) (t_2)	$Al_{0.36}Ga_{0.64}N$ Thickness (nm) (t_3)	Deposition Temperature (°C)
1	200	0.7	21	
2	200	1.2	21	
3	1000	0.7	21	1210
4	1000	0.7	31	
5	1000	1.2	31	

After the deposition, the sheet resistance, mobility, and charge were measured using Hall measurement using the Van der Pauw method. A capacitance–voltage (CV) measurement using a mercury CV tool was also performed to determine the charge control in the AlGaN/AlN/GaN heterostructure. The surface morphology was analyzed using atomic force microscopy (AFM) measurements with a Bruker Icon AFM in tapping mode. For analyzing the composition and thickness of the AlGaN barrier and AlN interlayer, omega-2theta and reciprocal space map (RSM) scans were performed on calibration structures using high-resolution XRD Panalytical Empyrean. X-ray reflectivity (XRR) measurements using Panalytical Empyrean were performed to measure the thickness of the AlGaN barrier layers.

3. Results and Discussion

Initially, three different TEGa GaN channel deposition temperatures, 1100 °C, 1210 °C, and 1270 °C, were studied, maintaining the UID GaN deposition temperature using TMGa at 1210 °C. The TEGa GaN channel thickness was kept at 70 nm to obtain the deposition rate from in situ reflectance using Laytec EpiTT. It was observed that, with increasing temperature from 1100 °C to 1270 °C, the deposition rate of GaN continued decreasing from 0.097 nm/s to 0.013 nm/s (Figure 2a). A sharp decrease in deposition rate from 0.083 nm/s to 0.013 nm/s was noticed between 1210°C and 1270 °C, suggesting a desorption-limited deposition regime at 1270 °C. This can affect the overall yield and increase the cost of manufacturing. The surface roughness increased by almost 38% and 102% for 1100 °C and 1270 °C, respectively, compared to 1210 °C (Figure 2b). However, the effect of surface roughness was different at 1100 °C and 1270 °C deposition temperatures. The step bunching was extremely high at 1100 °C, but the surface undulations were lower, whereas at 1270 °C the surface undulations were high but step bunching was minimal (Figure 2b). The possible reason behind the different types of surface roughness is mostly related to deposition kinetics. At 1100 °C, the incorporation of gallium in GaN was high (Figure 2a), but due to lower temperature, the lateral diffusion of the adatoms was affected, causing step bunching. On the contrary, at 1270 °C the Ga incorporation was extremely low (Figure 2a) (almost desorption limited deposition), but the presence of high thermal energy reduces step bunching. Also, increased undulations in the surface indicate a non-uniform deposition of GaN at 1270 °C. The most optimized deposition temperature was found to be 1210 °C, where the surface roughness was minimal but the deposition rate was moderate.

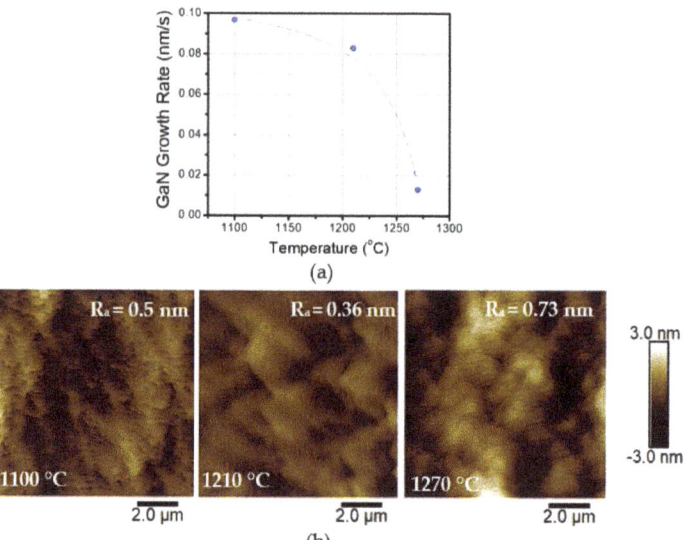

Figure 2. (a) Deposition rate (blue circles are experimentally obtained data, dashed blue line guides to the eyes) and FWHM of the omega–rocking curve and (b) surface roughness of GaN channel at different deposition temperatures.

The thickness of AlGaN was measured using an omega-2theta scan (Figure 3a) and the composition of the AlGaN layer was determined using the reciprocal space mapping (RSM) technique (Figure 3b). For the calibration of AlGaN composition and thickness, an AlGaN/GaN super-lattice (SL) structure (repetition 5–10) was deposited with 2–4 nm of AlGaN and 2–5 nm of GaN with different TMAl flows and a fixed TEGa flow rate (15 µmol/min). The GaN deposition rate was determined from the reflectance data obtained from the in–situ reflectance monitoring system during the thick (>100 nm) GaN layer deposition with the same TEGa flow rate. Next, the omega-2theta measurements of the AlGaN/GaN SL layers were performed using XRD and fitted using Panalytical X'pert Epitaxy software to obtain the thickness and composition. Three separate SL structures with three different TMAl flow rates along with a fixed TEGa flow rate were used to obtain a linear relationship between AlGaN composition, thickness, and the TMAl flow rate. After that, a specific TMAl flow rate (4.6 µmol/min) was chosen to obtain 36% AlGaN composition with a TEGa flow rate of 15 µmol/min. The thickness and composition values were further verified via XRR and RSM measurements, respectively. The fittings were performed using Panalytical AMASS 1.0 [34] and X'pert Epitaxy 4.5a [35] software for XRR and RSM measurements, respectively. The measured deposition rate of the AlGaN layer was near 0.14 nm/s. The RSM measured across the GaN (−1–14) reflection shows an almost fully strained 31 nm AlGaN (36%) barrier deposited on GaN with a 0.7 nm AlN interlayer. It also indicates that the AlGaN peak broadening was very low, which signifies high material quality. The AlN layer was deposited using TMAl and NH_3 under the same deposition conditions. The thickness (deposition rate 0.05 nm/s) of the AlN layer was determined using a similar calibration method to that described above.

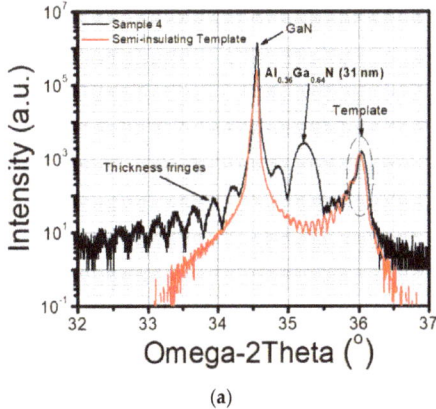
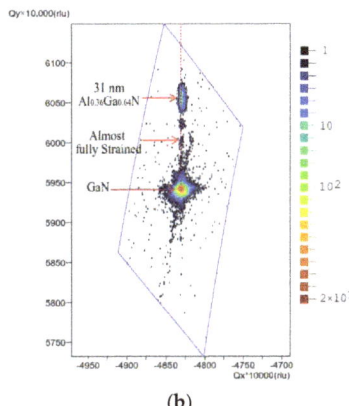

Figure 3. (a) Omega−2theta and (b) RSM scan of sample 4 at GaN (−1–14) reflection, measured using HRXRD.

Hall measurements were performed on all the samples as feedback for deposition condition optimizations. The sheet charge density and room temperature mobility of the samples are plotted in Figure 4a. The AlN thickness (t_2) was increased from 0.7 nm to 1.2 nm in sample 2 compared to sample 1 to understand the effect of alloy scattering. The mobility of sample 2 increased compared to sample 1 from 1110 cm^2/V·s to 1340 cm^2/V·s. So, the improved mobility signifies a reduction in alloy scattering with increasing AlN thickness. However, the sheet charge decreased slightly from $1.2 \times 10^{13}/cm^2$ to $1.12 \times 10^{13}/cm^2$, which is an outlier. The possible reason behind the slight decrease in the 2DEG charge density is the variation between the Fe-doped S.I. GaN templates, where the template used for the deposition of sample 2 had 30% higher surface roughness than the template of sample 1. This is also possibly due to the small variation in deposition uniformity from sample 1 to sample 2. In sample 3, the thickness of the UID-GaN layer (t_1) was increased from 200 nm to 1000 nm, which caused a significant change in mobility from 1110 cm^2/V·s (sample 3) to 1800 cm^2/V·s, and the charge also increased from $1.2 \times 10^{13}/cm^2$ (sample 3) to $1.33 \times 10^{13}/cm^2$. The possible reason behind the increase in mobility and charge is the increase in the distance between the 2DEG channel and the interface between UID-GaN (TMGa) and the semi-insulating GaN template. The semi-insulating layer is doped with Fe, which traps the unintentional background carriers in the GaN layer. If the 2DEG channel is in close proximity to the Fe-doped semi-insulating layer, then the electron transport might be affected due to the bulk-trapping phenomenon [36–38]. *Dmitri S. Arteev* et al. have also reported the effect of UID-GaN buffer or channel layer thickness on the properties of 2DEG charge and mobility in an AlGaN/AlN/GaN HEMT with a Fe-doped GaN buffer [39]. An increased AlGaN thickness (t_3) from 21 nm to 31 nm in sample 4 compared to sample 3 increased the sheet charge from $1.33 \times 10^{13}/cm^2$ (sample 3) to $1.46 \times 10^{13}/cm^2$ due to a reduction in surface depletion, while the mobility reduced from 1800 cm^2/V·s to 1710 cm^2/V·s compared to sample 3. The most likely reason for this is strain-induced mobility reduction in the channel [4] or increased carrier–carrier scattering [5]. Finally, in sample 5, we tried to reduce the alloy scattering by increasing the thickness of AlN (t_2) from 0.7 nm to 1.2 nm while keeping the AlGaN thickness (t_3) at 31 nm; this degraded the mobility from 1710 cm^2/V·s to 907 cm^2/V·s. However, it increased the sheet charge density from 1.46×10^{13}/cm^2 to 1.63×10^{13}/cm^2 due to a thicker AlN interlayer, which increased the apparent conduction band offset (ΔE_C). Figure 4b shows the overall sheet resistances of different samples. CV measurement was performed on samples 1–4 for analysis of the charge profile (Figure 4c) with different thicknesses of the UID-GaN (t_1), AlN interlayer (t_2), and AlGaN barrier (t_3) in the AlGaN/AlN/GaN HEMT structures. The CV measurement on sample 5 was corrupted as it had micro-cracks (not shown in Figure 4c). The plateau of the capacitance profile in Figure 4c indicates the presence of 2DEG in the AlGaN/AlN/GaN heterostructure. An increase in AlN interlayer thickness (t_2) from 0.7 nm (sample 1) to 1.2 nm (sample 1) shifted the pinch-off voltage from −5.3 V to −6.13 V, which indicates an increase in 2DEG sheet charge density in sample 2. This is expected with the increase in AlN interlayer thickness. However, Hall measurement showed a lowering in the 2DEG charge density from 1.2×10^{13}/cm^2 to 1.12×10^{13}/cm^2 in sample 2 compared to sample 1. This difference between the Hall and CV measurements might be related to the variations in surface roughness and/or deposition uniformity. Also, an increase in the thickness of the UID-GaN layer (t_1) in sample 3 from 200 nm to 1000 nm showed a reduction in the pinch-off voltage by −1.35 V, indicating an increase in the 2DEG charge density. The increase in the 2DEG charge is associated with a reduction in the unintentional incorporation of acceptor-type Fe dopants in the UID-GaN channel from the Fe-doped GaN buffer layer [39]. The pinch-off voltage further decreased to −8.85 V (sample 4) from −6.65 V (sample 3) when the AlGaN barrier thickness was increased from 21 nm to 31 nm, indicating an enhancement in the 2DEG charge density. At low temperatures (77 K), sample 3 and sample 4 show Hall mobilities of 8570 cm^2/V·s and 7830 cm^2/V·s (Table 2), which are comparable (>5500 cm^2/V·s) with the state-of-the-art (<300 Ω/□) low-temperature mobilities recorded for thick and high-composition AlGaN barrier HEMT structures [8,13,26], which proves that the AlN/GaN interface is extremely smooth and the material quality in the channel is significantly high (Figure 5).

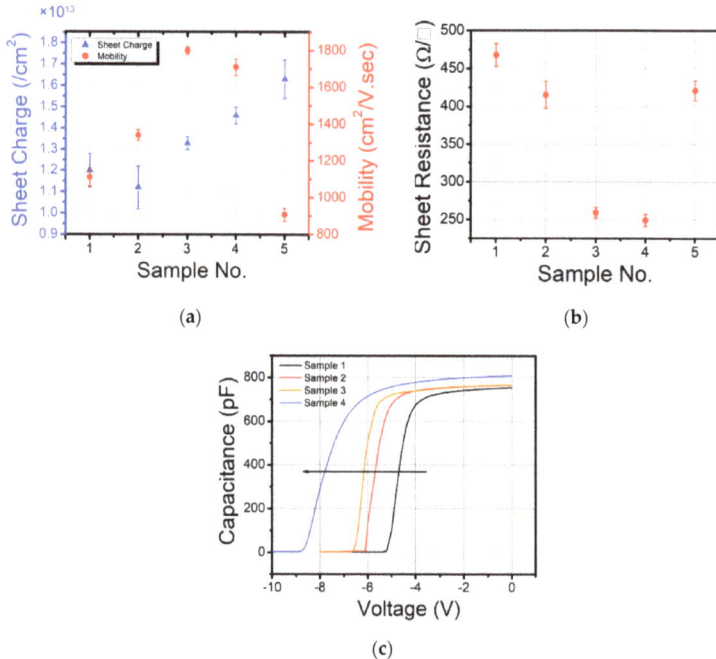

Figure 4. (a) 2DEG sheet charge and mobility, (b) sheet resistance, and (c) capacitance–voltage measurement of different samples.

Table 2. Comparative data between sample 3, sample 4, and sample 5.

Sample No.	Hall Measurement				R_a (nm)
	n_s (cm^{-2}) × 10^{13}	μ (cm^2/V·s) (300 K)	μ (cm^2/V·s) (77 K)	R_{SH} (Ω/□)	
3	1.33	1800	8570	259	0.42
4	1.46	1710	7830	249	0.27
5	1.63	907	3840	421	0.44

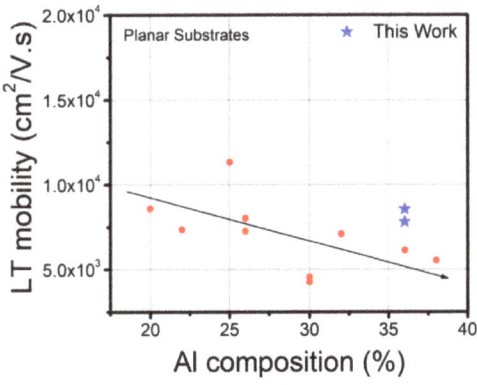

Figure 5. Comparison of low-temperature (LT) (77 K) mobility of 2DEG of AlGaN/GaN HEMT with respect to aluminum composition on planar substrates; references [8,13,15,27,40–44].

Comparing samples 4 (t_2 = 0.7 nm) and 5 (t_2 = 1.2 nm), the degradation in RT and LT mobility from 1710 cm^2/V·s to 907 cm^2/V·s and 7830 cm^2/V·s to 3840 cm^2/V·s, respectively, could be analyzed partly with AFM scans. In Figure 6a,b, the AFM scans (10 μm × 10 μm) of sample 4 and sample 5 are displayed. The surface roughness increased in sample 5 by 63% compared to sample 4 and more step bunching appeared. Also, Figure 6c shows that there were micro-cracks present in sample 5. Therefore, with a higher AlN thickness (1.2 nm) along with a thicker AlGaN (t_1 = 31 nm) layer, the strain in the barrier layer became extremely high and enough to generate micro-cracks in sample 5. So, the degradation in 2DEG mobility in the channel was possibly due to increased interface roughness and therefore a degradation in the interface quality [26,45,46]. The large degradation of the 2DEG mobility in sample 5 can be understood from Figure 6c. Micro-cracks increased the scattering of the 2DEG significantly, thus decreasing the mobility. Further optimization of the deposition process can indeed improve the 2DEG mobility.

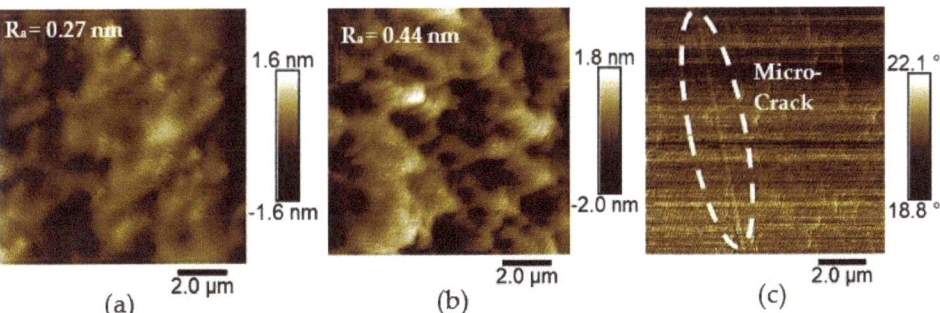

Figure 6. AFM height sensor scans of (**a**) sample 4 and (**b**) sample 5, and (**c**) phase sensor scan of sample 5 showing micro−crack.

The XRD omega-rocking curve measurement was performed on sample 4 and sample 5 to analyze the quality of the AlGaN barrier. Figure 7a,b show the omega-rocking curve of the (002) and (105) AlGaN planes, respectively, on sample 4 and sample 5. The increased AlN thickness (t_2 = 1.2 nm) of sample 5 increased the FWHM of the AlGaN layer by 25 arc-sec in the (002) plane and 395 arc-sec in the (105) plane compared to sample 4 with AlN thickness t_2 = 0.7 nm. The dislocation density in the AlGaN layer tended to increase with increasing strain, thus degrading the interface quality and reducing the 2DEG mobility significantly.

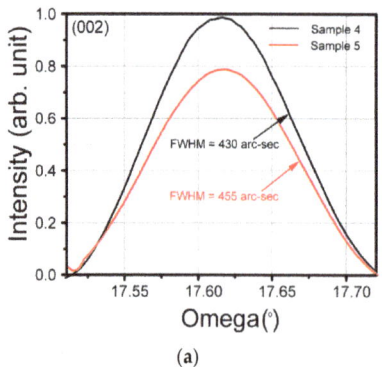
(a)

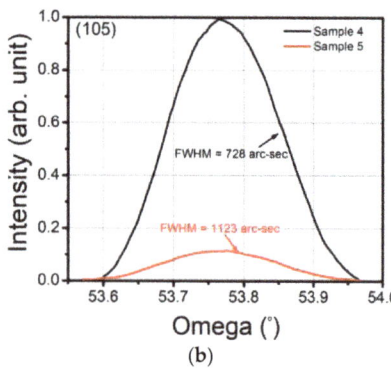
(b)

Figure 7. Omega-rocking curve of the AlGaN (**a**) (002) plane and (**b**) (105) plane of sample 4, and sample 5 with 31 nm $Al_{0.36}Ga_{0.64}N$/AlN/GaN HEMT.

From the above analyses, it can be seen that to achieve a thick (>30 nm) high-composition (>35%) crack-free AlGaN barrier layer HEMT with very low sheet resistance (<250 Ω/□), the deposition process needs to be optimized quite significantly. The experimental data prove that it is possible to grow high-quality and high-power AlGaN/AlN/GaN HEMT structures on sapphire with very low sheet resistance.

Figure 8 shows the transfer length method (TLM) measurement of the sheet resistance and contact resistance of 0.52 Ω.mm and sheet resistance of 248 Ω/□ of sample 3, which is similar to the sheet resistance obtained from the Hall measurement. The small difference between the Hall measurement and the TLM measurement might come from different contact-formation procedures, where indium contacts were used for the Hall measurement and Ti/Al/Ni contacts were deposited using the e-beam evaporation technique for the measurement of TLM. Also, the Hall measurement may give an average value of the sheet resistance, whereas the TLM measurement provides localized sheet resistance.

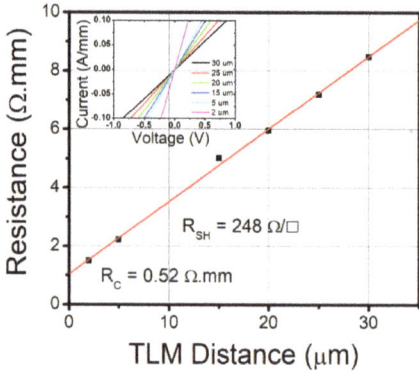

Figure 8. TLM measurement of sample 3, with 21 nm $Al_{0.36}Ga_{0.64}N$/AlN/GaN HEMT structure.

Figure 9a shows the importance of this work. Most of the AlGaN/GaN HEMTs that have been reported previously with sheet resistance of ~250 Ω/□ or less either have a thinner barrier layer (≤25 nm) or lower Al composition (≤30%). It is clear from Figure 9a that obtaining low sheet resistance with a thicker (>30 nm) and high-composition (>35%) AlGaN barrier is important. The combination of high barrier thickness, high composition, and low sheet resistance will ensure high-power operation with reduced gate leakage and increased breakdown. Figure 9b shows that the high-performance HEMTs were mostly deposited on SiC substrates. However, in this work, we demonstrate a state-of-the-art and simplified AlGaN/AlN/GaN HEMT design on sapphire that can operate at a similar level to a GaN HEMT deposited on SiC or GaN substrates that are relatively costlier.

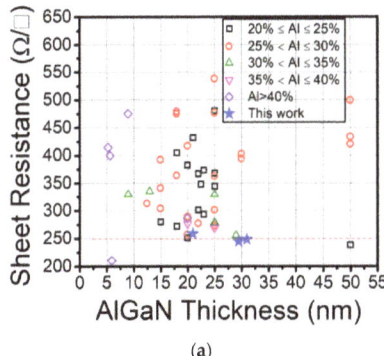
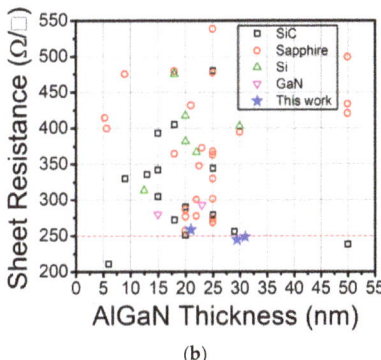

Figure 9. Sheet resistance variation with AlGaN thickness for (**a**) different Al compositions and (**b**) different substrates; references [4,7,8,12–15,26,27,40–44,47–69].

4. Conclusions

In conclusion, a high-quality thick-barrier $Al_{0.36}Ga_{0.64}N/AlN/GaN$ HEMT structure on sapphire with state-of-the-art sheet resistance has been deposited with the help of TEGa and a controlled-deposition process. High mobility is recorded at both room temperature and 77 K, proving the low interface roughness between the barrier and channel. A thick and high-composition crack-free AlGaN barrier HEMT was achieved using a deposition rate below 0.15 nm/s. Large spontaneous and piezoelectric polarization-induced charges can be obtained in the channel by using a high-composition (>35%) and thick (>30 nm) AlGaN barrier, which helps to reduce the sheet resistance. In conclusion, the experimental data show the potential of Ga-polar AlGaN/AlN/GaN HEMT on sapphire substrates, which can not only handle higher power but can be designed to be cost-effective too.

Author Contributions: Conceptualization, S.M., C.G. and S.S.P.; methodology, S.M., C.G. and S.S.P.; validation, S.M., C.G. and S.S.P.; formal analysis, S.M., C.L., C.G. and S.S.P.; investigation, S.M., C.G. and S.S.P.; resources, C.G. and S.S.P.; data curation, S.M., C.L., J.C., M.T.A., S.S., R.B. and G.W.; writing—original draft preparation, S.M.; writing—review and editing, S.M., C.G. and S.S.P.; visualization, S.M., C.G. and S.S.P.; supervision, C.G. and S.S.P.; project administration, C.G. and S.S.P.; funding acquisition, C.G. and S.S.P. All authors have read and agreed to the published version of the manuscript.

Funding: This work was funded by the Office of Naval Research: N00014-22-1-2267 and monitored by Paul Maki.

Data Availability Statement: The data that support the findings of this study are available from the corresponding authors upon reasonable request.

Conflicts of Interest: The authors declare no conflict of interest.

References

1. Shen, L.; Palacios, T.; Poblenz, C.; Corrion, A.; Chakraborty, A.; Fichtenbaum, N.; Keller, S.; DenBaars, S.P.; Speck, J.S.; Mishra, U.K. Unpassivated High Power Deeply Recessed GaN HEMTs with Fluorine-Plasma Surface Treatment. *IEEE Electron Device Lett.* **2006**, *27*, 214–216. [CrossRef]
2. Higashiwaki, M.; Mimura, T.; Matsui, T. AlGaN/GaN Heterostructure Field-Effect Transistors on 4H-SiC Substrates with Current-Gain Cutoff Frequency of 190 GHz. *Appl. Phys. Express* **2008**, *1*, 021103. [CrossRef]
3. Wright, A.; Nelson, J.M. Consistent Structural Properties for AlN, GaN, and InN. *Phys. Rev. B* **1995**, *51*, 7866–7869. [CrossRef]
4. Polyakov, V.M.; Cimalla, V.; Lebedev, V.; Köhler, K.; Müller, S.; Waltereit, P.; Ambacher, O. Impact of al Content on Transport Properties of Two-Dimensional Electron Gas in GaN/Al$_x$Ga$_{1-X}$N/GaN Heterostructures. *Appl. Phys. Lett.* **2010**, *97*, 142112. [CrossRef]
5. Ahmadi, E.; Keller, S.; Mishra, U.K. Model to Explain the Behavior of 2DEG Mobility with Respect to Charge Density in N-Polar and Ga-Polar AlGaN-GaN Heterostructures. *J. Appl. Phys.* **2016**, *120*, 115302. [CrossRef]
6. Yamada, A.; Junya, Y.; Nakamura, N.; Kotani, J. Low-Sheet-Resistance High-Electron-Mobility Transistor Structures with Strain-Controlled High-Al-Composition AlGaN Barrier Grown by MOVPE. *J. Cryst. Growth* **2021**, *560*, 126046. [CrossRef]
7. Wang, X.; Hu, G.; Ma, Z.; Ran, J.; Wang, C.; Xiao, H.; Tang, J.; Li, J.; Wang, J.; Zeng, Y.; et al. AlGaN/AlN/GaN/SiC HEMT Structure with High Mobility GaN Thin Layer as Channel Grown by MOCVD. *J. Cryst. Growth* **2007**, *298*, 835–839. [CrossRef]
8. Gaska, R.; Yang, J.; Andrei, O.; Chen, Q.Y.; Ijaz Khan, M.; Orlov, A.; Snider, G.L.; Shur, M. Electron Transport in AlGaN–GaN Heterostructures Grown on 6H–SiC Substrates. *Appl. Phys. Lett.* **1998**, *72*, 707–709. [CrossRef]
9. Melton, W.A.; Pankove, J.I. GaN Growth on Sapphire. *J. Cryst. Growth* **1997**, *178*, 168–173. [CrossRef]
10. O'Hanlon, T.J.; Zhu, T.; Massabuau, F.; Oliver, R.A. Dislocations at Coalescence Boundaries in Heteroepitaxial GaN/Sapphire Studied after the Epitaxial Layer Has Completely Coalesced. *Ultramicroscopy* **2021**, *231*, 113258. [CrossRef]
11. Cao, Y.; Wang, K.; Orlov, A.O.; Xing, H.G.; Jena, D. Very Low Sheet Resistance and Shubnikov–De-Haas Oscillations in Two-Dimensional Electron Gases at Ultrathin Binary AlN/GaN Heterojunctions. *Appl. Phys. Lett.* **2008**, *92*, 152112. [CrossRef]
12. Palacios, T.; Suh, C.-S.; Chakraborty, A.; Keller, S.; DenBaars, S.P.; Mishra, U.K. High-Performance E-Mode AlGaN/GaN HEMTs. *IEEE Electron Device Lett.* **2006**, *27*, 428–430. [CrossRef]
13. Ding, G.-J.; Guo, L.; Zhang, X.; Chen, Y.; Xu, P.; Jia, H.; Zhou, J.; Chen, H. Characterization of Different-Al-Content AlGaN/GaN Heterostructures on Sapphire. *Sci. China Phys. Mech. Astron.* **2010**, *53*, 49–53. [CrossRef]
14. Palacios, T.; Chakraborty, A.; Rajan, S.; Poblenz, C.; Keller, S.; DenBaars, S.P.; Speck, J.S.; Mishra, U.K. High-Power AlGaN/GaN HEMTs for Ka-Band Applications. *IEEE Electron Device Lett.* **2005**, *26*, 781–783. [CrossRef]
15. Gutierrez, P.; Tomas, A. *Optimization of the High Frequency Performance of Nitride-Based Transistors—Order No. 3206406*; University of California: Santa Barbara, CA, USA, 2006.
16. Xiao, M.; Ma, Y.; Cheng, K.; Liu, K.; Xie, A.; Beam, E.; Cao, Y.; Zhang, Y. 3.3 KV Multi-Channel AlGaN/GaN Schottky Barrier Diodes with P-GaN Termination. *IEEE Electron Device Lett.* **2020**, *41*, 1177–1180. [CrossRef]
17. Nela, L.; Erine, C.; Ma, J.; Yildirim, H.K.; van Erp, R.; Peng, X.; Cheng, K.; Matioli, E. High-Performance Enhancement-Mode AlGaN/GaN Multi-Channel Power Transistors. In Proceedings of the 2021 33rd International Symposium on Power Semiconductor Devices and ICs (ISPSD), Nagoya, Japan, 30 May–3 June 2021; pp. 143–146. [CrossRef]
18. Jiang, H.; Zhu, R.; Lyu, Q.; Lau, K.M. High-Voltage P-GaN HEMTs with OFF-State Blocking Capability after Gate Breakdown. *IEEE Electron Device Lett.* **2019**, *40*, 530–533. [CrossRef]
19. Jiang, H.; Lyu, Q.; Zhu, R.; Xiang, P.; Cheng, K.; Lau, K.M. 1300 v Normally-off P-GaN Gate HEMTs on Si with High ON-State Drain Current. *IEEE Trans. Electron Devices* **2021**, *68*, 653–657. [CrossRef]
20. Posthuma, N.; You, S.; Stoffels, S.; Wellekens, D.; Liang, H.; Zhao, M.; De Jaeger, B.; Geens, K.; Ronchi, N.; Decoutere, S.; et al. An Industry-Ready 200 Mm P-GaN E-Mode GaN-On-Si Power Technology. In Proceedings of the 2018 IEEE 30th International Symposium on Power Semiconductor Devices and ICs (ISPSD), Chicago, IL, USA, 13–17 May 2018. [CrossRef]
21. Lee, J.-H.; Ju, J.J.; Atmaca, G.; Kim, J.-G.; Kang, S.-H.; Lee, Y.S.; Lee, S.-H.; Lim, J.-W.; Kwon, H.-S.; Lisesivdin, S.B.; et al. High Figure-of-Merit (V^2_{BR}/R_{ON}) AlGaN/GaN Power HEMT With Periodically C-Doped GaN Buffer and AlGaN Back Barrier. *IEEE J. Electron Devices Soc.* **2018**, *6*, 1179–1186. [CrossRef]
22. Bahat-Treidel, E.; Brunner, F.; Hilt, O.; Cho, E.J.; Würfl, J.; Tränkle, G. AlGaN/GaN/GaN:C Back-Barrier HFETs With Breakdown Voltage of Over 1 kV and Low $R_{ON} \times A$. *IEEE Trans. Electron Devices* **2010**, *57*, 3050–3058. [CrossRef]
23. Hao, J.; Xing, Y.; Fu, K.; Zhang, P.; Song, L.; Chen, F.; Yang, T.; Deng, X.; Zhang, S.; Zhang, B. Influence of Channel/Back-Barrier Thickness on the Breakdown of AlGaN/GaN MIS-HEMTs. *J. Semicond.* **2018**, *39*, 094003. [CrossRef]
24. Wang, H.Y.; Chiu, H.C.; Hsu, W.C.; Liu, C.M.; Chuang, C.Y.; Liu, J.Z.; Huang, Y.L. The Impact of Al$_x$Ga$_{1-x}$N Back Barrier in AlGaN/GaN High Electron Mobility Transistors (HEMTs) on Six-Inch MCZ Si Substrate. *Coatings* **2020**, *10*, 570. [CrossRef]
25. Einfeldt, S.; Kirchner, V.; Heinke, H.; Dießelberg, M.; Figge, S.; Vogeler, K.; Hommel, D. Strain Relaxation in AlGaN under Tensile Plane Stress. *J. Appl. Phys.* **2000**, *88*, 7029–7036. [CrossRef]
26. Wang, C.; Wang, X.; Hu, G.; Wang, J.; Xiao, H.; Li, J. The Effect of AlN Growth Time on the Electrical Properties of Al$_{0.38}$Ga$_{0.62}$N/AlN/GaN HEMT Structures. *J. Cryst. Growth* **2006**, *289*, 415–418. [CrossRef]
27. Wang, C.; Wang, X.; Hu, G.; Wang, J.; Li, J.; Wang, Z. Influence of AlN Interfacial Layer on Electrical Properties of High-Al-Content Al$_{0.45}$Ga$_{0.55}$N/GaN HEMT Structure. *Appl. Surf. Sci.* **2006**, *253*, 762–765. [CrossRef]

28. Heikman, S.; Keller, S.; DenBaars, S.P.; Mishra, U.K. Growth of Fe Doped Semi-Insulating GaN by Metalorganic Chemical Vapor Deposition. *Appl. Phys. Lett.* **2002**, *81*, 439–441. [CrossRef]
29. Heikman, S.; Keller, S.; Mates, T.; DenBaars, S.P.; Mishra, U.K. Growth and Characteristics of Fe-Doped GaN. *J. Cryst. Growth* **2003**, *248*, 513–517. [CrossRef]
30. Saxler, A.; Walker, D.; Kung, P.; Zhang, X.; Manijeh, R.; Solomon, J.S.; Mitchel, W.C.; Vydyanath, H.R. Comparison of Trimethyl-gallium and Triethylgallium for the Growth of GaN. *Appl. Phys. Lett.* **1997**, *71*, 3272–3274. [CrossRef]
31. Kwang, H.S.; Dong, J.K.; Moon, Y.-T.; Park, S.-J. Characteristics of GaN Grown by Metalorganic Chemical Vapor Deposition Using Trimethylgallium and Triethylgallium. *J. Cryst. Growth* **2001**, *233*, 439–445. [CrossRef]
32. Hospodková, A.; František, H.; Tomáš, H.; Zuzana, G.; Hubik, P.; Hývl, M.; Pangrác, J.; Dominec, F.; Košutová, T. Electron Transport Properties in High Electron Mobility Transistor Structures Improved by V-Pit Formation on the AlGaN/GaN Interface. *ACS Appl. Mater. Interfaces* **2023**, *15*, 19646–19652. [CrossRef]
33. Stringfellow, G.B. Novel Precursors for Organometallic Vapor Phase Epitaxy. *J. Cryst. Growth* **1993**, *128*, 503–510. [CrossRef]
34. *Panalytical AMASS, 1.0*, Malvern Panalytical B.V.: Almelo, The Netherlands, 2018.
35. Panalytical X'pert Epitaxy, 4. *Panalytical X'pert Epitaxy, 4.5a*, Malvern Panalytical B.V.: Almelo, The Netherlands, 2018.
36. Li, X.; Bergsten, J.; Nilsson, D.; Örjan, D.; Pedersen, H.; Niklas, R.; Janzén, E.; Forsberg, U. Carbon Doped GaN Buffer Layer Using Propane for High Electron Mobility Transistor Applications: Growth and Device Results. *Appl. Phys. Lett.* **2015**, *107*, 262105. [CrossRef]
37. Zanato, D.; Gokden, S.; Balkan, N.; Ridley, B.K.; Schaff, W.J. The Effect of Interface-Roughness and Dislocation Scattering on Low Temperature Mobility of 2D Electron Gas in GaN/AlGaN. *Semicond. Sci. Technol.* **2004**, *19*, 427–432. [CrossRef]
38. Liu, J.P.; Ryou, J.-H.; Yoo, D.; Zhang, Y.; Limb, J.; Horne, C.A.; Shen, S.-C.; Dupuis, R.D.; Hanser, A.; Preble, E.A.; et al. III-Nitride Heterostructure Field-Effect Transistors Grown on Semi-Insulating GaN Substrate without Regrowth Interface Charge. *Appl. Phys. Lett.* **2008**, *92*, 133513. [CrossRef]
39. Arteev, D.S.; Sakharov, A.V.; Lundin, W.V.; Zavarin, E.E.; Nikolaev, A.E.; Tsatsulnikov, A.F.; Ustinov, V.M. Scattering Analysis of AlGaN/AlN/GaN Heterostructures with Fe-Doped GaN Buffer. *Materials* **2022**, *15*, 8945. [CrossRef]
40. Bergsten, J.; Chen, J.-T.; Gustafsson, S.; Malmros, A.; Forsberg, U.; Thorsell, M.; Janzén, E.; Rorsman, N. Performance Enhancement of Microwave GaN HEMTs without an AlN-Exclusion Layer Using an Optimized AlGaN/GaN Interface Growth Process. *IEEE Trans. Electron Devices* **2016**, *63*, 333–338. [CrossRef]
41. Hu, W.; Ma, B.; Li, D.; Narukawa, M.; Miyake, H.; Hiramatsu, K. Mobility Enhancement of 2DEG in MOVPE-Grown AlGaN/AlN/GaN HEMT Structure Using Vicinal (0 0 0 1) Sapphire. *Superlattices Microstruct.* **2009**, *46*, 812–816. [CrossRef]
42. Aggerstam, T.; Lourdudoss, S.; Radamson, H.H.; Sjödin, M.; Lorenzini, P.; Look, D.C. Investigation of the Interface Properties of MOVPE Grown AlGaN/GaN High Electron Mobility Transistor (HEMT) Structures on Sapphire. *Thin Solid Film.* **2006**, *515*, 705–707. [CrossRef]
43. Miyoshi, M.; Egawa, T.; Ishikawa, H. Study on Mobility Enhancement in MOVPE-Grown AlGaN/AlN/GaN HEMT Structures Using a Thin AlN Interfacial Layer. *Solid-State Electron.* **2006**, *50*, 1515–1521. [CrossRef]
44. Wang, X.L.; Wang, C.M.; Hu, G.X.; Wang, J.X.; Chen, T.S.; Jiao, G.; Li, J.P.; Zeng, Y.P.; Li, J.M. Improved DC and RF Performance of AlGaN/GaN HEMTs Grown by MOCVD on Sapphire Substrates. *Solid-State Electron.* **2005**, *49*, 1387–1390. [CrossRef]
45. Elsass, C.R.; Poblenz, C.; Heying, B.; Fini, P.; Petroff, P.M.; DenBaars, S.P.; Mishra, U.K.; Speck, J.S.; Saxler, A.; Elhamrib, S.; et al. Influence of Growth Temperature and Thickness of AlGaN Caps on Electron Transport in AlGaN/GaN Heterostructures Grown by Plasma-Assisted Molecular Beam Epitaxy. *Jpn. J. Appl. Phys.* **2001**, *40*, 6235. [CrossRef]
46. Keller, S.L.; Parish, G.; Fini, P.T.; Heikman, S.; Chen, C.; Zhang, N.; DenBaars, S.P.; Mishra, U.K.; Wu, Y. Metalorganic Chemical Vapor Deposition of High Mobility AlGaN/GaN Heterostructures. *J. Appl. Phys.* **1999**, *86*, 5850–5857. [CrossRef]
47. Gustafsson, S.; Chen, T., Jr.; Bergsten, J.; Forsberg, U.; Thorsell, M.; Janzen, E.; Rorsman, N. Dispersive Effects in Microwave AlGaN/AlN/GaN HEMTs with Carbon-Doped Buffer. *IEEE Trans. Electron Devices* **2015**, *62*, 2162–2169. [CrossRef]
48. Kumar, V.; Lu, W.; Schwindt, R.; Kuliev, A.; Simin, G.; Yang, J.; Khan, M.A.; Adesida, I. AlGaN/GaN HEMTs on SiC with F/Sub T/ of over 120 GHz. *IEEE Electron Device Lett.* **2002**, *23*, 455–457. [CrossRef]
49. Yuen, Y.W.; Chiu, Y.-S.; Luong, T.-T.; Lin, T.-M.; Yen, T.H.; Lin, Y.C.; Chang, E.Y. Growth and Fabrication of AlGaN/GaN HEMT on SiC Substrate. In Proceedings of the 10th IEEE International Conference on Semiconductor Electronics (ICSE), Kuala Lumpur, Malaysia, 19–21 September 2012; pp. 729–732. [CrossRef]
50. Hao, Y.; Yang, L.; Ma, X.; Ma, J.; Cao, M.; Pan, C.; Wang, C.; Zhang, J. High-Performance Microwave Gate-Recessed AlGaN/AlN/GaN MOS-HEMT with 73% Power-Added Efficiency. *IEEE Electron Device Lett.* **2011**, *32*, 626–628. [CrossRef]
51. Shen, L.; Heikman, S.; Moran, B.; Coffie, R.; Zhang, N.-Q.; Buttari, D.; Smorchkova, I.P.; Keller, S.; DenBaars, S.P.; Mishra, U.K. AlGaN/AlN/GaN High-Power Microwave HEMT. *IEEE Electron Device Lett.* **2001**, *22*, 457–459. [CrossRef]
52. Yusuke, K.; Ozaki, S.; Okamoto, N.; Hara, N.; Takao, O. Low-Resistance and Low-Thermal-Budget Ohmic Contact by Introducing Periodic Microstructures for AlGaN/AlN/GaN HEMTs. *IEEE Trans. Electron Devices* **2022**, *69*, 3073–3078. [CrossRef]
53. Palacios, T.; Chakraborty, A.; Heikman, S.; Keller, S.; DenBaars, S.P.; Mishra, U.K. AlGaN/GaN High Electron Mobility Transistors with InGaN Back-Barriers. *IEEE Electron Device Lett.* **2006**, *27*, 13–15. [CrossRef]
54. Gong, J.-M.; Wang, Q.; Yan, J.-D.; Liu, F.-Q.; Feng, C.; Wang, X.-L.; Wang, Z.-G. Comparison of GaN/AlGaN/AlN/GaN HEMTs Grown on Sapphire with Fe-Modulation-Doped and Unintentionally Doped GaN Buffer: Material Growth and Device Fabrication. *Chin. Phys. Lett.* **2016**, *33*, 117303. [CrossRef]

55. Zhang, H.; Sun, Y.; Song, K.; Xing, C.; Yang, L.; Wang, D.; Yu, H.; Xiang, X.; Gao, N.; Xu, G.; et al. Demonstration of AlGaN/GaN HEMTs on Vicinal Sapphire Substrates with Large Misoriented Angles. *Appl. Phys. Lett.* **2021**, *119*, 072104. [CrossRef]
56. Mahaboob, I.; Yakimov, M.M.; Hogan, K.; Rocco, E.; Tozier, S.; Shahedipour-Sandvik, F. Dynamic Control of AlGaN/GaN HEMT Characteristics by Implementation of a P-GaN Body-Diode-Based Back-Gate. *IEEE J. Electron Devices Soc.* **2019**, *7*, 581–588. [CrossRef]
57. Wang, X.; Wang, C.; Hu, G.; Wang, J.; Li, J. Room Temperature Mobility above 2100 cm^2/vs in $Al_{0.3}Ga_{0.7}N$/AlN/GaN Heterostructures Grown on Sapphire Substrates by MOCVD. *Phys. Status Solidi C* **2006**, *3*, 607–610. [CrossRef]
58. Miyoshi, M.; Imanishi, A.; Egawa, T.; Ishikawa, H.; Asai, K.; Shibata, T.; Tanaka, M.; Oda, O. DC Characteristics in High-Quality AlGaN/AlN/GaN High-Electron-Mobility Transistors Grown on AlN/Sapphire Templates. *Jpn. J. Appl. Phys.* **2005**, *44*, 6490–6494. [CrossRef]
59. Cai, Y.; Zhou, Y.; Chen, K.J.; Lau, K.M. High-Performance Enhancement-Mode AlGaN/GaN HEMTs Using Fluoride-Based Plasma Treatment. *IEEE Electron Device Lett.* **2005**, *26*, 435–437. [CrossRef]
60. Li, H.; Keller, S.; DenBaars, S.P.; Mishra, U.K. Improved Properties of High-Al-Composition AlGaN/GaN High Electron Mobility Transistor Structures with Thin GaN Cap Layers. *Jpn. J. Appl. Phys.* **2014**, *53*, 095504. [CrossRef]
61. Zhang, K.; Chen, X.; Mi, M.; Zhao, S.; Chen, Y.; Zhang, J.; Ma, X.; Hao, Y. Enhancement-Mode AlGaN/GaN HEMTs with Thin and High al Composition Barrier Layers Using O_2 Plasma Implantation. *Phys. Status Solidi A* **2014**, *212*, 1081–1085. [CrossRef]
62. Cheng, J.; Yang, X.; Sang, L.; Guo, L.; Hu, A.; Xu, F.; Tang, N.; Wang, X.; Shen, B. High Mobility AlGaN/GaN Heterostructures Grown on Si Substrates Using a Large Lattice-Mismatch Induced Stress Control Technology. *Appl. Phys. Lett.* **2015**, *106*, 142106. [CrossRef]
63. Ubukata, A.; Yano, Y.; Shimamura, H.; Yamaguchi, A.; Tabuchi, T.; Matsumoto, K. High-Growth-Rate AlGaN Buffer Layers and Atmospheric-Pressure Growth of Low-Carbon GaN for AlGaN/GaN HEMT on the 6-in.-Diameter Si Substrate Metal-Organic Vapor Phase Epitaxy System. *J. Cryst. Growth* **2013**, *370*, 269–272. [CrossRef]
64. Zhang, J.; He, L.; Li, L.; Ni, Y.; Que, T.; Liu, Z.; Wang, W.; Zheng, J.; Huang, Y.; Chen, J.; et al. High-Mobility Normally off Al_2O_3/AlGaN/GaN MISFET with Damage-Free Recessed-Gate Structure. *IEEE Electron Device Lett.* **2018**, *39*, 1720–1723. [CrossRef]
65. Bouzid-Driad, S.; Maher, H.; Defrance, N.; Hoel, V.; De Jaeger, J.-C.; Renvoise, M.; Frijlink, P. AlGaN/GaN HEMTs on Silicon Substrate with 206-GHz F_{Max}. *IEEE Electron Device Lett.* **2013**, *34*, 36–38. [CrossRef]
66. Latrach, S.; Frayssinet, E.; Defrance, N.; Chenot, S.; Cordier, Y.; Gaquière, C.; Maaref, H. Trap States Analysis in AlGaN/AlN/GaN and InAlN/AlN/GaN High Electron Mobility Transistors. *Curr. Appl. Phys.* **2017**, *17*, 1601–1608. [CrossRef]
67. Xu, X.; Zhong, J.; So, H.; Norvilas, A.; Sommerhalter, C.; Senesky, D.G.; Tang, M. Wafer-Level MOCVD Growth of AlGaN/GaN-On-Si HEMT Structures with Ultra-High Room Temperature 2DEG Mobility. *AIP Adv.* **2016**, *6*, 115016. [CrossRef]
68. Chu, J.; Wang, Q.; Jiang, L.; Feng, C.; Li, W.; Liu, H.; Xiao, H.; Wang, X. Room Temperature 2DEG Mobility above 2350 cm^2/V·s in AlGaN/GaN HEMT Grown on GaN Substrate. *J. Electron. Mater.* **2021**, *50*, 2630–2636. [CrossRef]
69. Yu, H.; Alian, A.; Peralagu, U.; Zhao, M.; Waldron, N.; Parvais, B.; Collaert, N. Surface State Spectrum of AlGaN/AlN/GaN Extracted from Static Equilibrium Electrostatics. *IEEE Trans. Electron Devices* **2021**, *68*, 5559–5564. [CrossRef]

Disclaimer/Publisher's Note: The statements, opinions and data contained in all publications are solely those of the individual author(s) and contributor(s) and not of MDPI and/or the editor(s). MDPI and/or the editor(s) disclaim responsibility for any injury to people or property resulting from any ideas, methods, instructions or products referred to in the content.

Article

First Demonstration of Extrinsic C-Doped Semi-Insulating N-Polar GaN Using Propane Precursor Grown on Miscut Sapphire Substrate by MOCVD

Swarnav Mukhopadhyay *, Surjava Sanyal, Guangying Wang, Chirag Gupta and Shubhra S. Pasayat

Electrical & Computer Engineering, University of Wisconsin-Madison, Madison, WI 53706, USA; ssanyal2@wisc.edu (S.S.); gwang265@wisc.edu (G.W.); cgupta9@wisc.edu (C.G.); shubhra@ece.wisc.edu (S.S.P.)
* Correspondence: swarnav.mukhopadhyay@wisc.edu

Abstract: In this study, carbon-doped semi-insulating N-polar GaN on a sapphire substrate was prepared using a propane precursor. Controlling the deposition rate of N-polar GaN helped to improve the carbon incorporation efficiency, providing a semi-insulating behavior. The material quality and surface roughness of the N-polar GaN improved with modified deposition conditions. C-doping using 1.8 mmol/min of propane gave an abrupt doping profile near the GaN/sapphire interface, which was useful for obtaining semi-insulating N-polar GaN grown on sapphire. This study shows that further development of the deposition process will allow for improved material quality and produce a state-of-the-art N-polar semi-insulating GaN layer.

Keywords: N-polar GaN; propane doping; semi-insulating N-polar GaN; C-doped N-polar GaN; growth rate optimization

1. Introduction

N-polar GaN-based high-electron-mobility transistors (HEMTs) show great potential as high-power (>5 W/mm) and high-frequency (W-band, 94 GHz) devices [1–3]. A parasitic conduction path other than the two-dimensional electron gas (2DEG) and high buffer leakage can potentially restrict the high-power operation of these transistors. A semi-insulating (S.I.) buffer layer is necessary for HEMT devices to increase the breakdown voltage while preventing leakage current and RF loss. The presence of residual oxygen in the deposition chamber, acting as a shallow donor, makes it non-trivial to achieve a semi-insulating (S.I.) GaN layer via the metal-organic chemical vapor deposition (MOCVD) deposition process [4]. In N-polar GaN, the incorporation of oxygen (O) is almost two orders of magnitude higher than that in Ga-polar GaN, which makes it challenging to obtain semi-insulating behavior [5–9]. Some methods for growing an S.I. GaN layer involve doping with iron (Fe), manganese (Mn), magnesium (Mg), and carbon (C), acting as a deep-level compensating acceptor state within GaN [10–15].

The gettering process is also one of the ways to reduce the background O concentration by trapping it, but it requires ion implantation, which can degrade the material quality [16,17], increase the cost, and reduce the throughput during production [18]. So, this process is not used in this study; rather, in situ techniques such as doping are chosen for improved material quality and reduced cost of production.

Fe doping using ferrocene is widely accepted as a way of generating a deep-level compensating acceptor state for growing S.I. GaN buffer layers [19]. Omega-rocking curve XRD measurements from prior studies showed the high-quality Fe-doped (1.3×10^{19}/cm^3) S.I. Ga-polar GaN layer [19]. The full-width at half maximum (FWHM) of the omega-rocking curve was determined to be 253 and 481 arc-sec across (002) and (102) orientations of GaN, respectively [19]. The memory effect of Fe-doping in Ga-polar GaN is a well-known phenomenon (similar to that of Mn [12] and Mg [20] doping in Ga-polar GaN), where the doping profile of Fe cannot be turned off abruptly [19,21]. After switching off the Fe precursor during the deposition of GaN, the Fe tail (slow-turn-off) extends till 0.8–1 µm [19,21]. The memory effect of Fe doping limits its applicability in HEMTs as Fe can also act as deep-level buffer traps that can potentially cause dispersive or current collapse effects in high-power RF amplifiers [10,19]. Heikman et al. showed a slow solid-phase incorporation effect of Fe-doping in Ga-polar GaN [21] and, because of that, the compensation of unintentional oxygen (O) incorporation at GaN/sapphire interface becomes non-trivial [22]. This problem becomes worse due to the higher-temperature nucleation of N-polar GaN compared to Ga-polar GaN [23], where the O concentration at the GaN/sapphire interface becomes larger compared to Ga-polar GaN. Alternatively, C-doping can help to reduce the background carrier concentration (O) while maintaining an abrupt doping profile, which is beneficial for HEMTs. C-doping can be achieved by the intrinsic doping method by utilizing the methyl (CH_3) or ethyl (C_2H_5) group from the metal-organic precursors of gallium, such as Trimethyl-gallium (TMGA) or Triethyl-gallium (TEGa). Intrinsic C-doping can be obtained using modified process parameters such as deposition temperature, flow rate of TMGa or TEGa, reactor pressure, and V/III ratio. Another method for C-doping is extrinsic doping, which uses hydrocarbon precursors during deposition [24]. However, the most optimized deposition conditions for efficient intrinsic doping typically result in a 40–67% increase in the dislocation/defect density [10,24,25]. Hence, hydrocarbons like propane, ethylene, and iso-butane are favored as C-dopants, providing the flexibility to utilize optimized GaN deposition conditions while maintaining precise control over extrinsic carbon incorporation [24].

Shan Wu et al. [26] reported that background hydrogen [H] can affect the electrical properties of intrinsically C-doped Ga-polar GaN by forming C-H complexes. However, the same group has also shown that extrinsic C-doping using propane attracts 75% less H, thereby reducing the chance of the formation of C-H complexes [27]. Additionally, *Fichtenbaum* reported a background H concentration of $1.5 \times 10^{17}/cm^3$ in an N-polar GaN film grown on a miscut sapphire substrate [28]. It exhibits a deposition rate (~50 nm/min) similar to that in this study. So, in this study, only the C and O concentrations in the N-polar GaN were measured using the SIMS, as the H concentration in propane-doped semi-insulating N-polar GaN might not have significantly influenced the material and electrical properties compared to the concentrations of C and O.

In this study, we used propane as a source of carbon precursor for obtaining the S.I. N-polar GaN layer on the miscut sapphire substrate using a metal–organic chemical vapor deposition (MOCVD) reactor. As per the authors' knowledge, this is the first attempt to demonstrate a C-doped S.I. The N-polar GaN layer with propane as a hydrocarbon precursor showed minimal deterioration (10–20% higher FWHM of the omega-rocking curve compared to the state-of-the-art curve) of the crystalline properties [29]. The incorporation efficiency of propane was improved in the N-polar GaN epitaxial layer under optimized deposition conditions. The main purpose of this study was to show the incorporation of carbon into N-polar GaN using propane as a gaseous precursor.

2. Experimental Methods

Propane-doped N-polar GaN was deposited on a miscut sapphire substrate using the MOCVD deposition technique. Trimethylgallium (TMGa) and ammonia (NH_3) were used as group-III and group-V precursors, respectively. Propane was used as a carbon (C) dopant, and H_2 was used as a carrier gas. The miscut sapphire sample was heated at 1300 °C, and nitridation was performed prior to the deposition of a high-temperature (1145 °C) nucleation layer at a reactor chamber pressure of 100 mbar and a V/III ratio of ~8000, similar to that mentioned by Keller et al. [23]. After that, 50 nm unintentionally doped (UID) N-polar GaN followed by 1.75 µm thick UID or propane-doped N-polar GaN was deposited at a high temperature (HT) of 1270 °C. Two different deposition conditions were used for the propane doping experiments in the 1.75 µm thick UID or propane-doped layer, where the TMGa molar flow was varied to observe the propane incorporation. For the first deposition condition (G_1), 137 µmol/min TMGa was used with propane concentrations of 0 to 3.6 mmol/min. The second deposition condition (G_2) had an increased TMGa molar flow of 204 µmol/min, and the propane concentration varied from 0 to 2.4 mmol/min. The propane flow did not increase to 3.6 mmol/min with G_2 as semi-insulating behavior was obtained before 2.4 mmol/min. The V/III ratio was fixed for both G_1 and G_2. The deposition conditions and corresponding propane concentrations used for G_1 and G_2 are listed in Table 1.

Table 1. Deposition conditions along with the propane flow rate applied for the deposition of N-polar GaN.

Deposition Condition	TMGa Flow Rate (µmol/Min)	Propane Flow Rate (mmol/Min)
G_1	137	0
		1.2
		2.4
		3.6
G_2	204	0
		0.6
		1.2
		1.4
		1.8
		2.4

The background electron concentration of all the samples was determined by Hall measurement using the Van der Pauw geometry. The omega-rocking curve was measured using a Panalytical Empyrean high-resolution X-ray diffraction (XRD) tool (Malvern Panalytical, Malvern, UK) for analyzing the material quality. The surface roughness of the deposited samples was measured using a Bruker Icon atomic force microscope (AFM) (Bruker, Billerica, USA). Secondary ion-mass spectroscopy (SIMS) measurement was performed using a Cameca instrument (Cameca, Gennevilliers, France) for measuring the O and C concentrations in the samples using Cs^+ ions. All the samples showed hillock-free atomically smooth surfaces under a Nomarski optical microscope (Nikon, Minato, Japan).

3. Results and Discussion

The carbon incorporation efficiency was primarily determined using Hall measurements. The first deposition condition (G_1) with a TMGa flow rate of 137 µmol/min showed an exponential decay in the background carrier concentration with an increasing propane flow rate (Figure 1a). It was observed in the literature that the carbon concentration in Ga-polar GaN linearly depends on the propane flow rate [30]. So, it was expected that with increasing propane flow, the background carrier concentration would decrease and the compensation ratio $\left(CR = \frac{n_{UID} - n_{C-doped}}{n_{UID}} \times 100\right)$ would increase linearly. It was also determined from our experiment in N-polar GaN that the CR followed an almost linear profile with the propane flow rate (Figure 1b). At a very high propane flow rate, there was a small deviation in the CR from linearity. This might be related to the increased dislocation density, as shown in Figure 2. An increase in the dislocation density can increase the background O incorporation, as observed by Szymanski et al. [9].

Omega-rocking curve measurements indicated that with increasing propane flow rate, the off-axis (102) full-width half maxima (FWHM) increased, suggesting an increased edge dislocation density (Figure 2). The on-axis (002) FWHM remained nearly the same. The surface roughness appeared to be more or less similar with an increasing propane flow rate (Figure 3), and the step flow deposition was not affected by the increased propane flow. The RMS roughness of these propane-doped N-polar GaN samples was found to be less than 1.5 nm. Next, in the G_2 series, the optimized deposition condition was used to increase the C-incorporation efficiency, such that even with the use of a lower amount of propane, S.I. behavior could be obtained.

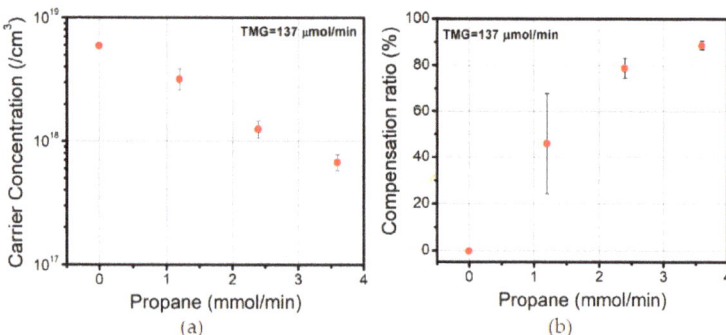

Figure 1. (a) Background carrier concentration and (b) compensation ratio of N-polar GaN deposited using G_1 condition at different propane flow rates.

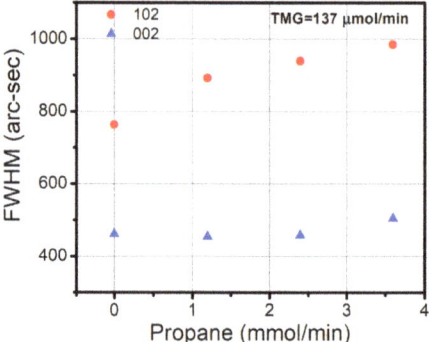

Figure 2. FWHM of the omega-rocking curve, which was measured using HRXRD on propane-doped N-polar GaN samples grown using G_1 deposition condition.

To improve the incorporation of C using a propane precursor, the G_2 deposition condition was used in the following studies. According to Lundin et al. [25], an increased deposition rate of Ga-polar GaN helps to increase the incorporation efficiency of C into GaN. This phenomenon was also studied in this work to understand the behavior of propane incorporation in N-polar GaN. Increasing the TMGa flow rate from 137 to 204 µmol/min increased the deposition rate from 60 to 90 nm/min and showed not only higher propane incorporation but also a reduced background carrier concentration. This might be related to the trapping of more carbon atoms during the deposition while limiting C desorption and fewer chances for O incorporation due to the higher supersaturation condition [9,31]. Hall measurements showed S.I. behavior with propane concentration > 1.8 mmol/min (Figure 4a). An exponential decay in the background carrier concentration was observed with monotonically increasing propane flow. Simultaneously, the slope of the CR with respect to the propane flow increased significantly for G_2 (Figure 4b), indicating a better incorporation efficiency of propane at a higher TMGa molar flow rate.

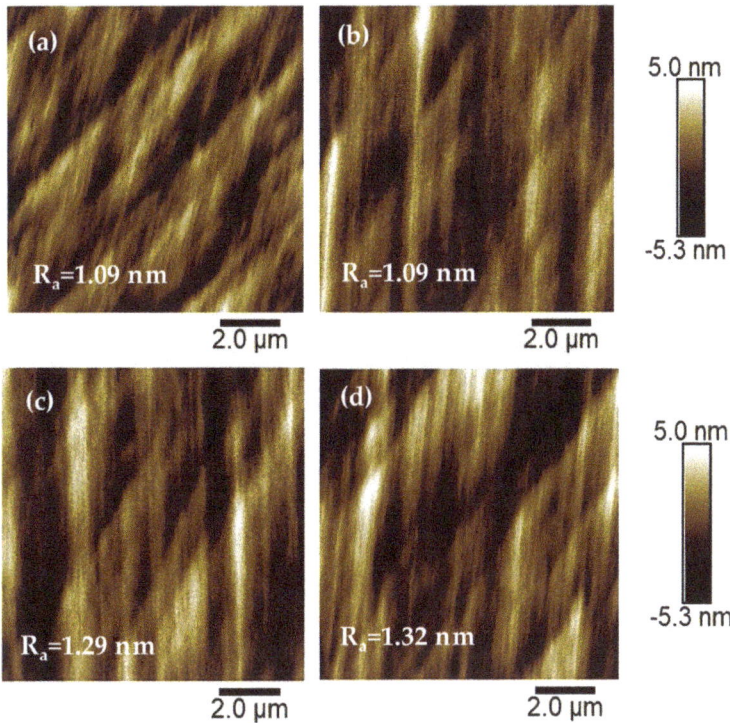

Figure 3. AFM scans (10 μm × 10 μm) of N-polar GaN samples growth under G_1 deposition condition with propane concentration of (**a**) 0, (**b**) 1.2, (**c**) 2.4, and (**d**) 3.6 mmol/min.

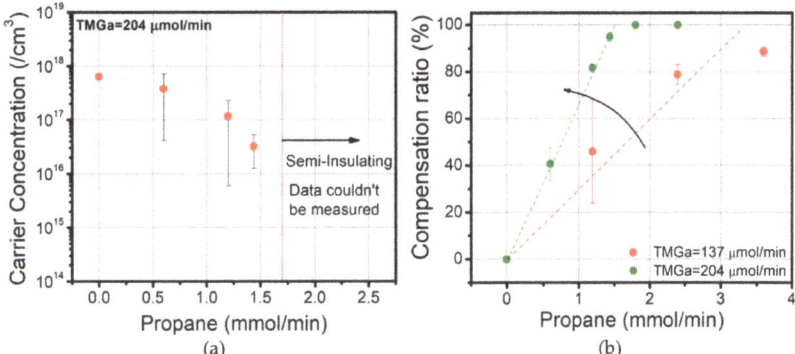

Figure 4. (**a**) Background carrier concentration and (**b**) compensation ratio of N-polar GaN at different propane flow rates.

The omega-rocking curve measurement of the G_2 sample showed similar trends as those of G_1. The FWHM increased with an increasing propane flow rate. Although it followed a comparable trend, it exhibited notably lower (10–20%) FWHM values in contrast to G_1, suggesting a reduced dislocation density in comparison to G_1 (Figure 5). The implementation of the modified deposition condition G_2 assisted in achieving a semi-insulating N-polar GaN material while utilizing a decreased propane flow rate. It improved the material's crystalline quality while reducing the background carrier concentration, indicating that the compensation of O mainly occurs due to the increased incorporation efficiency of C atoms and not due to the increase in dislocation and defect density. The FWHM value of the semi-insulating C-doped N-polar GaN was still higher than that of N-polar UID GaN. However, Lundin et al. demonstrated that fully optimized deposition conditions in heavily (2×10^{19}/cm^3) propane-doped Ga-polar semi-insulating GaN can provide a very low FWHM of the omega-rocking curve of the (002) and (102) planes of ~250 arc-sec and ~320 arc-sec, respectively [25]. So, further development by optimizing the growth conditions, such as V/III ratio, growth rate of N-polar GaN, growth temperature, and reactor pressure, needs to be performed to improve the FWHM parameter of semi-insulating N-polar GaN.

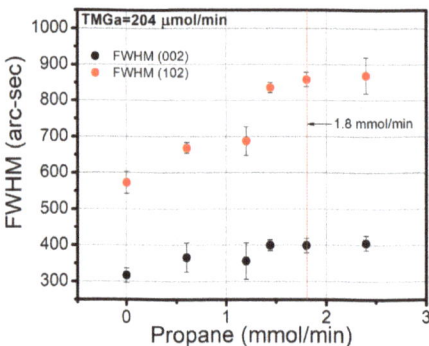

Figure 5. FWHM of the omega-rocking curve, measured using HRXRD on propane-doped N-polar GaN samples grown using G_1 and G_2 deposition conditions.

The surface roughness remained below 1.5 nm for the 10×10 μm^2 AFM scan area, indicating device quality propane doped S.I. N-polar GaN samples using propane flow of 1.8 mmol/min and 2.4 mmol/min (Figure 6).

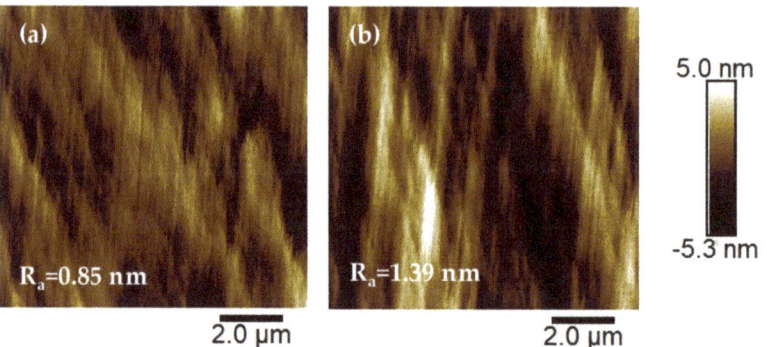

Figure 6. AFM scans (10 μm × 10 μm) of N-polar GaN samples deposited using G_2 deposition conditions at propane flow rates (**a**) 1.8 and (**b**) 2.4 mmol/min.

The SIMS measurement carried out for the G_2 deposition condition using 2.4 mmol/min propane flow shows more than one order of higher C-concentration compared to the O and Si impurity levels, demonstrating complete semi-insulating behavior (Figure 7). The detection levels of O, C, and Si are 10^{16}/cm^3, 5×10^{16}/cm^3, and 10^{16}/cm^3, respectively. Figure 7 also shows the abrupt doping profile of C at the GaN/sapphire interface, which is crucial for obtaining S.I. behavior. It is known from the literature that the background O level is almost 2–4 times higher at the GaN/sapphire interface compared to bulk GaN [32]. In order to compensate for the increased background O at the N-polar GaN/sapphire interface, the C concentration needs to be significantly high near the interface. In Figure 7, it can be observed that the doping concentration of C reaches nearly $2 \times 10^{18}/\text{cm}^3$ in the vicinity of the interface, effectively compensating the background O concentration. The increased O needs to be fully compensated for by C, or the region near the interface needs to be fully depleted to obtain the semi-insulating property of N-polar GaN. It is possible to achieve a semi-insulating property if the C concentration is sufficiently higher than the increased O concentration. One of the possible reasons for obtaining a semi-insulating behavior might be due to more than one order higher C concentration compared to the O concentration in the bulk N-polar GaN. This higher C concentration might be sufficient to fully compensate for the increased O level at the interface. Another possible reason might be that the higher O level is near the nucleation layer, and it is not fully electrically conductive. In the nucleation layer, the N-polar GaN film quality might not be as good as that of the bulk, which appears from the thin island growth method in the nucleation layer [23]. So, even though the O concentration is higher near the interface, it does not act like a fully conductive path. However, the resolution of the SIMS measurement tool is insufficient to capture the thickness (possibly less than 50 nm) of the N-polar GaN/sapphire interface, where elevated O levels might exist. To accurately understand the physics behind the compensation mechanism near the N-polar GaN/sapphire interface, an extensive study using energy-dispersive X-ray spectroscopy needs to be performed, which is currently beyond the scope of this study. As a result, obtaining an accurate estimation of the O concentration at the interface is challenging.

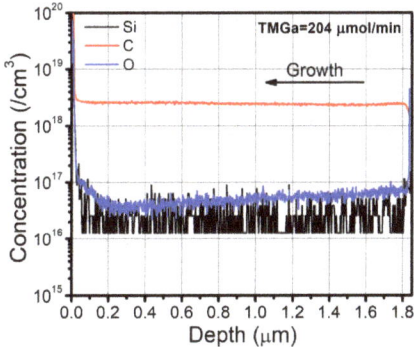

Figure 7. SIMS measurement of propane-doped N-polar GaN sample with a propane flow rate of 2.4 mmol/min and deposition condition G_2.

4. Conclusions

This study is the first demonstration of propane-doped semi-insulating N-polar GaN deposited on top of a sapphire substrate. It was observed that with a higher deposition rate of N-polar GaN, the C-incorporation efficiency increased. Improved surface morphology and material quality were obtained by increasing the TMGa molar flow rate while obtaining an S.I. behavior with a comparatively lower propane flow rate. A propane flow rate of 1.8 mmol/min resulted in S.I. N-polar GaN along with a minimal surface roughness of 0.85 nm, which was deposited using 204 µmol/min of TMGa. This signifies that further development of propane doping in N-polar GaN will lead to improved material quality and production of S.I. C-doped N-polar GaN with a controlled doping profile.

Author Contributions: Conceptualization, S.M., C.G. and S.S.P.; methodology, S.M., C.G. and S.S.P.; validation, S.M., C.G. and S.S.P.; formal analysis, S.M., C.G. and S.S.P.; investigation, S.M., C.G. and S.S.P.; resources, C.G. and S.S.P.; data curation, S.M., S.S. and G.W.; writing—original draft preparation, S.M.; writing—review and editing, S.M., C.G. and S.S.P.; visualization, S.M., C.G. and S.S.P.; supervision, C.G. and S.S.P.; project administration, C.G. and S.S.P.; funding acquisition, C.G. and S.S.P. All authors have read and agreed to the published version of the manuscript.

Funding: This study was funded by the Office of Naval Research: N00014-22-1-2267 and monitored by Paul Maki.

Data Availability Statement: The data that support the findings of this study are available from the corresponding authors upon reasonable request.

Conflicts of Interest: The authors declare no conflict of interest.

References

1. Romanczyk, B.; Li, W.; Guidry, M.; Hatui, N.; Krishna, A.; Wurm, C.; Keller, S.; Mishra, U.K. N-Polar GaN-on-Sapphire Deep Recess HEMTs With High W-Band Power Density. *IEEE Electron Device Lett.* **2020**, *41*, 1633–1636. [CrossRef]
2. Romanczyk, B.; Wienecke, S.; Guidry, M.; Li, H.; Ahmadi, E.; Zheng, X.; Keller, S.; Mishra, U.K. Demonstration of Constant 8 W/mm Power Density at 10, 30, and 94 GHz in State-of-the-Art Millimeter-Wave N-Polar GaN MISHEMTs. *IEEE Trans. Electron Devices* **2017**, *65*, 45–50. [CrossRef]
3. Li, W.; Romanczyk, B.; Guidry, M.; Akso, E.; Hatui, N.; Wurm, C.; Liu, W.; Shrestha, P.; Collins, H.; Clymore, C.; et al. Record RF Power Performance at 94 GHz From Millimeter-Wave N-Polar GaN-on-Sapphire Deep-Recess HEMTs. *IEEE Trans. Electron Devices* **2023**, *70*, 2075–2080. [CrossRef]
4. Chung, B.-C.; Gershenzon, M. The influence of oxygen on the electrical and optical properties of GaN crystals grown by metalorganic vapor phase epitaxy. *J. Appl. Phys.* **1992**, *72*, 651–659. [CrossRef]
5. Fichtenbaum, N.; Mates, T.; Keller, S.; DenBaars, S.; Mishra, U. Impurity incorporation in heteroepitaxial N-face and Ga-face GaN films grown by metalorganic chemical vapor deposition. *J. Cryst. Growth* **2008**, *310*, 1124–1131. [CrossRef]
6. Tanikawa, T.; Kuboya, S.; Matsuoka, T. Control of impurity concentration in N-polar ($000\bar{1}$) GaN grown by metalorganic vapor phase epitaxy. *Phys. Status Solidi (b)* **2017**, *254*, 1600751. [CrossRef]
7. Tavernier, P.; Margalith, T.; Williams, J.; Green, D.; Keller, S.; DenBaars, S.; Mishra, U.; Nakamura, S.; Clarke, D. The growth of N-face GaN by MOCVD: Effect of Mg, Si, and In. *J. Cryst. Growth* **2004**, *264*, 150–158. [CrossRef]
8. Sumiya, M.; Yoshimura, K.; Ohtsuka, K.; Fuke, S. Dependence of impurity incorporation on the polar direction of GaN film growth. *Appl. Phys. Lett.* **2000**, *76*, 2098–2100. [CrossRef]
9. Szymanski, D.; Wang, K.; Kaess, F.; Kirste, R.; Mita, S.; Reddy, P.; Sitar, Z.; Collazo, R. Systematic oxygen impurity reduction in smooth N-polar GaN by chemical potential control. *Semicond. Sci. Technol.* **2022**, *37*, 015005. [CrossRef]
10. Li, X.; Bergsten, J.; Nilsson, D.; Danielsson, Ö.; Pedersen, H.; Rorsman, N.; Janzén, E.; Forsberg, U. Carbon doped GaN buffer layer using propane for high electron mobility transistor applications: Growth and device results. *Appl. Phys. Lett.* **2015**, *107*, 262105. [CrossRef]
11. Amilusik, M.; Zajac, M.; Sochacki, T.; Lucznik, B.; Fijalkowski, M.; Iwinska, M.; Wlodarczyk, D.; Somakumar, A.K.; Suchocki, A.; Bockowski, M. Carbon and Manganese in Semi-Insulating Bulk GaN Crystals. *Materials* **2022**, *15*, 2379. [CrossRef]
12. Yamamoto, T.; Sazawa, H.; Nishikawa, N.; Kiuchi, M.; Ide, T.; Shimizu, M.; Inoue, T.; Hata, M. Reduction in Buffer Leakage Current with Mn-Doped GaN Buffer Layer Grown by Metal Organic Chemical Vapor Deposition. *Jpn. J. Appl. Phys.* **2013**, *52*, 08JN12. [CrossRef]
13. Reshchikov, M.A.; Ghimire, P.; Demchenko, D.O. Magnesium acceptor in gallium nitride. I. Photoluminescence from Mg-doped GaN. *Phys. Rev. B* **2018**, *97*, 205204. [CrossRef]
14. Hautakangas, S.; Oila, J.; Alatalo, M.; Saarinen, K.; Liszkay, L.; Seghier, D.; Gislason, H.P. Vacancy Defects as Compensating Centers in Mg-Doped GaN. *Phys. Rev. Lett.* **2003**, *90*, 137402. [CrossRef]
15. Nakamura, S.; Iwasa, N.; Senoh, M.S.M.; Mukai, T.M.T. Hole Compensation Mechanism of P-Type GaN Films. *Jpn. J. Appl. Phys.* **1992**, *31*, 1258. [CrossRef]
16. Ronning, C.; Carlson, E.P.; Thomson, D.B.; Davis, R.F. Optical activation of Be implanted into GaN. *Appl. Phys. Lett.* **1998**, *73*, 1622–1624. [CrossRef]
17. Kucheyev, S.; Williams, J.; Pearton, S. Ion implantation into GaN. *Mater. Sci. Eng. R Rep.* **2001**, *33*, 51–108. [CrossRef]
18. Majid, A.; Ali, A.; Zhu, J.; Wang, Y.; Yang, H. An evidence of defect gettering in GaN. *Phys. B Condens. Matter* **2008**, *403*, 2495–2499. [CrossRef]
19. Heikman, S.; Keller, S.; DenBaars, S.P.; Mishra, U.K. Growth of Fe doped semi-insulating GaN by metalorganic chemical vapor deposition. *Appl. Phys. Lett.* **2002**, *81*, 439–441. [CrossRef]
20. Xing, H.; Green, D.S.; Yu, H.; Mates, T.; Kozodoy, P.; Keller, S.; DenBaars, S.P.; Mishra, U.K. Memory Effect and Redistribution of Mg into Sequentially Regrown GaN Layer by Metalorganic Chemical Vapor Deposition. *Jpn. J. Appl. Phys.* **2003**, *42*, 50–53. [CrossRef]
21. Heikman, S.; Keller, S.; Mates, T.; DenBaars, S.; Mishra, U. Growth and characteristics of Fe-doped GaN. *J. Cryst. Growth* **2003**, *248*, 513–517. [CrossRef]
22. Heikman, S.J. MOCVD Growth Technologies for Applications in AlGaN/GaN High Electron Mobility Transistors. Ph.D. Thesis, University of California, Santa Barbara, CA, USA, September 2002.
23. Keller, S.; Li, H.; Laurent, M.; Hu, Y.; Pfaff, N.; Lu, J.; Brown, D.F.; A Fichtenbaum, N.; Speck, J.S.; DenBaars, S.P.; et al. Recent progress in metal-organic chemical vapor deposition of ($000\bar{1}$) N-Polar Group-III Nitrides. *Semicond. Sci. Technol.* **2014**, *29*, 113001. [CrossRef]

24. Li, X.; Danielsson, Ö.; Pedersen, H.; Janzén, E.; Forsberg, U. Precursors for carbon doping of GaN in chemical vapor deposition. *J. Vac. Sci. Technol. B* **2015**, *33*, 021208. [CrossRef]
25. Lundin, W.; Sakharov, A.; Zavarin, E.; Kazantsev, D.; Ber, B.; Yagovkina, M.; Brunkov, P.; Tsatsulnikov, A. Study of GaN doping with carbon from propane in a wide range of MOVPE conditions. *J. Cryst. Growth* **2016**, *449*, 108–113. [CrossRef]
26. Wu, S.; Yang, X.; Zhang, Q.; Shang, Q.; Huang, H.; Shen, J.; He, X.; Xu, F.; Wang, X.; Ge, W.; et al. Direct evidence of hydrogen interaction with carbon: C–H complex in semi-insulating GaN. *Appl. Phys. Lett.* **2020**, *116*, 262101. [CrossRef]
27. Wu, S.; Yang, X.; Wang, Z.; Ouyang, Z.; Huang, H.; Zhang, Q.; Shang, Q.; Shen, Z.; Xu, F.; Wang, X.; et al. Influence of intrinsic or extrinsic doping on charge state of carbon and its interaction with hydrogen in GaN. *Appl. Phys. Lett.* **2022**, *120*. [CrossRef]
28. Fichtenbaum, N.A. *Growth of Nitrogen-Face Gallium Nitride by MOCVD*; Order No. 3330425; University of California: Santa Barbara, CA, USA, 2008.
29. Bisi, D.; Romanczyk, B.; Liu, X.; Gupta, G.; Brown-Heft, T.; Birkhahn, R.; Lal, R.; Neufeld, C.J.; Keller, S.; Parikh, P.; et al. Commercially Available N-polar GaN HEMT Epitaxy for RF Applications. In Proceedings of the 2021 IEEE 8th Workshop on Wide Bandgap Power Devices and Applications (WiPDA), Redondo Beach, CA, USA, 7–11 November 2021; pp. 250–254. [CrossRef]
30. Lesnik, A.; Hoffmann, M.P.; Fariza, A.; Bläsing, J.; Witte, H.; Veit, P.; Hörich, F.; Berger, C.; Hennig, J.; Dadgar, A.; et al. Properties of C-doped GaN. *Phys. Status Solidi (b)* **2016**, *254*, 1600708. [CrossRef]
31. Seifert, W.; Franzheld, R.; Butter, E.; Sobotta, H.; Riede, V. On the origin of free carriers in high-conducting n-GaN. *Cryst. Res. Technol.* **1983**, *18*, 383–390. [CrossRef]
32. Sumner, J.; Das Bakshi, S.; Oliver, R.A.; Kappers, M.J.; Humphreys, C.J. Unintentional doping in GaN assessed by scanning capacitance microscopy. *Phys. Status Solidi (b)* **2008**, *245*, 896–898. [CrossRef]

Disclaimer/Publisher's Note: The statements, opinions and data contained in all publications are solely those of the individual author(s) and contributor(s) and not of MDPI and/or the editor(s). MDPI and/or the editor(s) disclaim responsibility for any injury to people or property resulting from any ideas, methods, instructions or products referred to in the content.

MDPI AG
Grosspeteranlage 5
4052 Basel
Switzerland
Tel.: +41 61 683 77 34

Crystals Editorial Office
E-mail: crystals@mdpi.com
www.mdpi.com/journal/crystals

Disclaimer/Publisher's Note: The statements, opinions and data contained in all publications are solely those of the individual author(s) and contributor(s) and not of MDPI and/or the editor(s). MDPI and/or the editor(s) disclaim responsibility for any injury to people or property resulting from any ideas, methods, instructions or products referred to in the content.

www.ingramcontent.com/pod-product-compliance
Lightning Source LLC
LaVergne TN
LVHW070042120526
838202LV00101B/414